How Science Points to God

Also by Gerard M. Verschuuren
from Sophia Institute Press:

Forty Anti-Catholic Lies
A Mythbusting Apologist
Sets the Record Straight

In the Beginning
A Catholic Scientist Explains
How God Made Earth Our Home

A Catholic Scientist Proves God Exists

Dr. Gerard M. Verschuuren

How Science Points to God

SOPHIA INSTITUTE PRESS
Manchester, New Hampshire

Cover by David Ferris Design.

Cover image: vector design (275517830) © Marina Sun / Shutterstock.

Sophia Institute Press
Box 5284, Manchester, NH 03108
1-800-888-9344

www.SophiaInstitute.com

Sophia Institute Press® is a registered trademark of Sophia Institute.

Paperback ISBN 978-1-64413-151-0

eBook ISBN 978-1-64413-152-7

Library of Congress Control Number: 2020938767

First printing

Dedicated to Peter Kreeft
whose clear, philosophical mind
has cleared many obstacles
to bring faith and science back together

Contents

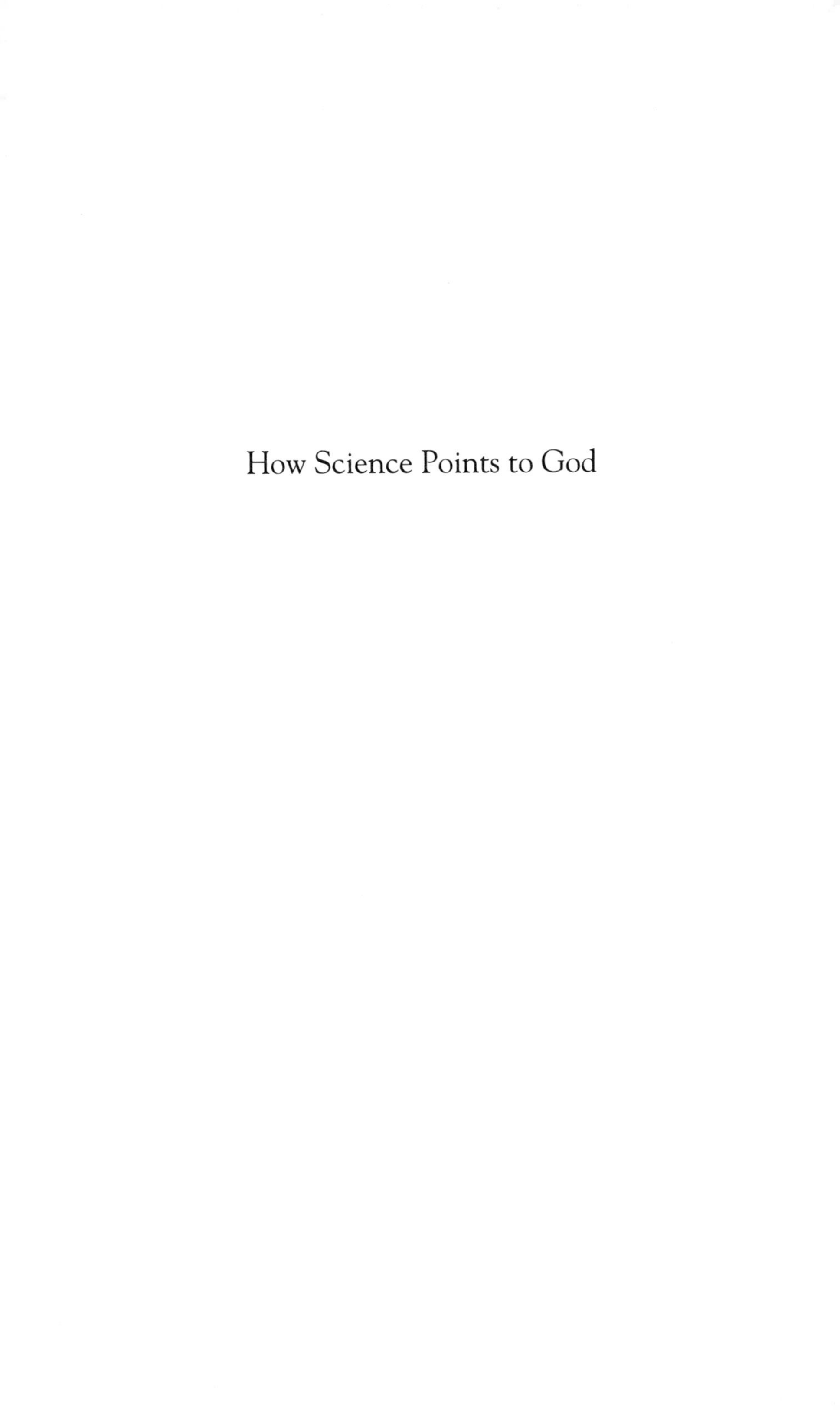

How Science Points to God

Preface

Does God exist? Where can we find an answer that everyone would accept? Could it be in science? Indeed, some of us like to make belief in God a scientific issue, but I have bad news for them: God is not a hypothesis that can be tested in the laboratory. Science cannot prove God's existence, but neither can it disprove the existence of God. Does that mean the existence of God is no longer worth our attention? Is that the end of religion?

Some of us do believe it is the end, making this question an all-or-nothing issue—either we choose science or we go for religion, leaving us no middle ground. But the situation is not that simple. The Boston College philosopher Peter Kreeft compares the situation with that Western in which one cowboy says to the other: "This town ain't big enough for both you and me. One of us has to leave."[1] Well, many nowadays like to declare science as the winner, so religion has to leave town. They replace the deep trust religious believers put in God with the total trust they place in science. There is not enough room for both.

[1] Peter Kreeft, "Pillars of Unbelief: Sartre," *National Catholic Register*, February 1988.

In this view, scientific expansion means religious withdrawal—so religion must be on its way out whenever science advances. But that is exactly where the misconception lies. Science doesn't gain territory at all when it makes new discoveries; it just learns more and more details about its own fixed and demarcated territory—which is the domain of all that can be dissected, counted, measured, and quantified. The rest is not part of its territory, but has been left for other "authorities" to handle.

This explains why, on a closer look, science in itself is not the problem for religion. Reality tells us that there are atheistic scientists as well as religious scientists. The former are no better scientists than the latter, nor vice versa. All of them are dedicated scientists who believe in the power of the scientific method. But they differ in one thing. The religious scientists keep their minds open and believe also in the power of religious faith, whereas the others close their minds to anything that cannot be dissected, counted, measured, or quantified. Seen this way, it is not science that "kills God," but rather particular scientists who do so in a very unscientific way.

So the quest for God remains on. Although science cannot prove or disprove God's existence, it still can and does point to God. We can find "hints" or "signs" of God's existence in what science has discovered. In this book, we will scan science and its findings for such pointers to God. They may not prove God's existence in the strict sense, but they certainly give us powerful indications that science could not exist if God did not exist.

1

How Scientific Assumptions Point to God

Without the assumption that there is an objectively real world, science would not be possible. Without the assumption that the world is comprehensible, science would not be possible. Without the assumption that our sensory perception is in essence reliable, science would not be possible. Without the assumption that our rationality is reliable; that there is order and uniformity in nature; that mathematics and logic are valid tools in reasoning, science would not be possible. Strange as it may sound, science is something you need to believe in before you can practice it.

Do scientists really hold these assumptions? Most philosophers—the professional skeptics among us—would say, "No, they don't." But when you ask scientists, their answer may be reluctant, but nevertheless affirmative when probed further. The reason why some of them may not immediately say, "Yes, we do," is that assumptions like these are usually hidden or unspoken, such that some scientists may not even realize they are holding them.

There are probably several reasons for this. One of them is that all people, scientists in particular, hold assumptions that are taken so completely for granted that they are never articulated. But another reason, if not the main reason, for this silence may be that each one of these assumptions seems to point to the existence of

God—and that's a threatening idea for some. If you just keep silent about these assumptions, they think, you neither have to admit you hold them nor explain why. Keeping silent would make your work as a scientist much easier. But how long can one remain silent?

Here are some of the assumptions scientists hold, or even must hold, whether they like to admit it or not.

The world is real for science

This assumption is about a world outside of us and independent of us, open to scientific discovery. This may seem an obvious assumption for scientists, but it has been questioned many times—questioned not only by some quantum physicists but also by skeptical philosophers such as David Hume and Immanuel Kant, who think we can't really know anything about the "real" world behind or beyond our observations. Some philosophers have even said that the entire world exists only as a dream in people's minds.

Should scientists take these skeptics seriously? Amusingly enough, scientists sometimes joke about their work with warnings such as, "Don't touch anything in a physics lab," or "Don't taste anything in a chemistry lab," or "Don't smell anything in a biology lab." But perhaps they should add this warning to the collection: "Don't trust anything in a philosophy department."

Scientists have indeed good reason not to trust those skeptical philosophical views. If there is no real world, then science becomes a hallucination too. So to keep from crumbling, science needs the assumption of a real world. Without the assumption of a real world outside of us and independent of us, there is no way anymore to distinguish facts from fictions, realities from illusions, and opinions from truths. If there is no objective truth, we are free to believe whatever we like, including utter nonsense. Once scientists give up on the assumption of a real outside world, and of the truths that come with it, they undermine their own findings, changing their science into mere science fiction.

No wonder the legendary Albert Einstein always protested vehemently against such an outcome: "The belief in an external world independent of the perceiving subject is the basis of all natural science."[2] Nevertheless, there have been some dissenting voices in the field of quantum physics. The work of quantum physicist Niels Bohr, for instance, seems to tell us that reality does not exist when we are not observing it. During one of their conversations, Bohr remembers that "Einstein suddenly stopped, turned to me and asked whether I really believed that the moon exists only when I look at it."[3]

What are we to make of this discussion? At least we can say that Bohr does acknowledge there is indeed a real world that we can observe, though only when we decide to do so. Besides, one should ask Bohr how we can measure something that doesn't exist, such that it must have come into existence at least before our measurement, and not because of our measurement.

Why does or should science still hold on to its truth claims? Well, if truth were at the mercy of some individuals, science would have to abandon all its universal claims. It is the real world that often forces scientists to revise their theories in order to come closer to the truth. To use a telling analogy: Who would ever want to drive across a bridge designed by engineers who believed their calculations were based merely on opinions instead of truths?

Besides, if the claim of these skeptics entails that our beliefs are mere artifacts or illusions, such a claim would act like a boomerang

[2] Albert Einstein, *Ideas and Opinions* (New York: Crown Publishers, 1954), 266.

[3] Abraham Pais, "Einstein and the quantum theory," *Reviews of Modern Physics* 51, no. 4 (October–December 1979): 907.

in destroying its own truth claims as well. This is an example of what G. K. Chesterton called "the suicide of thought."[4] To know is to know things in the real world, not to know mental abstractions in the mind. Those who say that the world of science is merely an illusion produced by our brains should be questioned seriously about this claim, as it would indeed be just an illusion by its own verdict. Their claim invalidates itself on the spot.

So this raises the question of how we know that what we know about the world is, in fact, about the real world. The shortest answer is: we don't. We can never prove that we're not all hallucinating, or collectively dreaming, or simply living in some kind of computer simulation. So we are dealing here with an assumption that science cannot prove on its own. Some people think the existence of mind-independent objects can be proven by giving a mighty kick to a stone. But a mighty kick is not proof. Objects perceived by our senses are not the same things as truly existing bodies—they may just be images on the retina or in the brain. Even in a dream, our experiences may resemble those of waking life, and yet they are illusions.

Therefore, there is no real proof that an objective external world does exist—for we could in fact all be living in a permanent dream world. Think of this: How would nearsighted people know that the world is not as blurred as they see it? Certainly not by comparing these blurry images with the "real" ones. Perhaps, though, corrective glasses may help them to see better what the real world looks like.

However, at the very moment we acknowledge that we can see better, we end up talking again about the existence of an objective, real world, distinct from our minds. That's where the buck stops. If there were no connection between what's out there and what's in our minds, then all we know or could know is our own

[4] G. K. Chesterton, *Orthodoxy* (New York: Doubleday, 1959), chap. 3.

thoughts. And since our minds have no connection with things, we could concoct whatever world we like. The bottom-line is this: our minds would no longer have any reality check. That would be detrimental to science.

How does this assumption of a real, objective world point to God? If there is indeed an objective world independent of our minds, then it is hard to avoid some crucial questions. Where does this objective world come from? Why is this world the way it is? Or the biggest question of all, as Gottfried Leibniz worded it: "Why is there something rather than nothing?"[5] Of course, we can leave such questions unanswered, but we cannot really deny or ignore them. Bertrand Russell, for instance, thought he could. He took a "brute fact" position when he said, "I should say that the universe is just there, and that's all."[6]

However, just to say that things are the way they are is not a very satisfying answer. Without any further explanation, it leaves us with a mere riddle of why there *is* something, why there *is* an objective world—that is, why something does exist at all. Nothing can explain its own existence. As Ludwig Wittgenstein put it, "Not *how* the world is, is the mystical, but *that* it is."[7] The "brute fact" position basically ignores the question instead of answering it.

The most reasonable answer to the above questions—indeed, the only satisfying one—is the one that points to God, who caused this external world to come into existence, just the way it is. God is the explanation why there is something rather than nothing.

[5] G. W. Leibniz, *Philosophical Essays*, trans. and ed. Roger Ariew and Daniel Garber (Indianapolis, IN: Hackett, 1989), 210.

[6] Transcript of the 1948 BBC radio debate between Russell and Fr. Frederick Copleston, S.J.

[7] Ludwig Wittgenstein, *Tractatus Logico-Philosophicus*, 6.44. Emphasis added.

Without God, there would most likely not be any objective world, actually nothing at all.

The world can be put to the test

Aristotle famously said, "All men by nature desire to know."[8] This desire starts early. Children constantly ask us "Why?" Aristotle goes on to note, "An indication of this is the delight we take in our senses." Translated: if you want to know, open your eyes and your ears. Science testifies to this. Scientists are masters of observation, investigation, and exploration by using their senses. Their work cannot be done from an armchair or behind a desk.

Scientists need to test their hypotheses, theories, and speculations by deriving test implications from them. If these test implications turn out to be true, then what scientists are testing *may* turn out to be true too—which is called confirmation. However, if these test implications turn out to be false, then what scientists are testing *must* turn out to be false, too—which is called falsification. For example, a test implication for the hypothesis that the earth is round would be that ships at the horizon disappear from sight. If that does indeed happen, then the hypothesis might be true.

Science is therefore an *empirical* enterprise. Aristotle would agree with this. But he would probably disagree that science is also *experimental*. That is, in a simple setting, scientists start with two variables, of which the independent variable is the one freely chosen and manipulated by the scientist during an experiment in order to trace its effect on the other, so-called dependent variable. Nowadays, scientists explore their field of study by using not only their senses but also their hands to manipulate their objects.

[8] Aristotle, *Physics* I, chap. 9, 192a16–19.

For modern minds, that is easily understood and actually expected. We no longer accept that science, like mathematics, should be done from behind a desk. And our experimental hands would be itching if they were not permitted to touch and manipulate nature. Nevertheless, centuries ago, the approach of most scholars was quite different.

In Aristotle's time, most Greek philosophers considered science to be theoretical knowledge—as distinct from the practical skill of people such as Archimedes. Their understanding was that knowledge about nature could not possibly be acquired by disturbing the delicate harmony of nature through what they considered "unnatural" interventions from without. In their eyes, an empirical approach would be incompatible with an experimental approach. Their reasoning went as follows: How can observation in an unnatural experiment be true to nature? By bending nature to one's own will, one could never discover its true features.

That is why many Greek scholars at the time detested what they saw as experimental interference. Experiments were not supposed to be part of "real science"[9]—at best, they belonged to the field of artisans, where technical skills were considered a form of art. Scholars usually remained distinct from artisans, with the rare exception of anatomists, who used lancets to remove what obstructed their view. By this technique, they acted as artisans to study as scholars. Aristotle used similar means during his rather accurate observations of the embryonic development of chickens. But in general, theoretical understanding and experimental intervention supposedly could not go together. An exception could be found in Alexandria, where Egyptian technicians and Greek theoreticians worked together.

[9] Originally, the term "science" (*scientia*) simply meant "knowledge." The term "scientist," in fact, did not exist until William Whewell coined the term in 1833.

It was Roger Bacon in the thirteenth century who, together with Robert Grosseteste and St. Albertus the Great, clearly articulated the "modern" conception of (natural) science. In doing so, they were the forerunners of the era in which so-called revolutionary science came to life (1300–1650). Roger Bacon introduced the distinction between "passive observation" as performed by the layman and "active experimentation" as done by the scientist. From then on, theoretical understanding and experimental intervention were supposed to go hand in hand. Experiments had become an essential part of the "new" science. The growth and popularity of this new movement were partly made possible by another Bacon: Francis Bacon. The "Baconian" ideal of science is based on the principle that science grows by gathering empirical as well as experimental data.

Due to the influence of Roger and Francis Bacon, all natural sciences became experimental, in principle, and the fact is that experiments cannot be performed without manipulation. Experiments depend on the notion that there is a knowable mechanism linking cause to effect. Experiments work by our exerting control over a cause and then noting its effects. They basically put nature to the test. Experimentation is no longer seen as something foreign or unnatural, but has become the hallmark of any scientific study.

How does the assumption that the world can be put to the test point to God? Belief in a Creator God entails that nature is not a divine but a created entity. Consequently, nature is not divine in itself; only its Maker is—which opens the door for scientific exploration. It is through scientific experiments that we can "read" God's Mind, so to speak. In contrast, if the world itself were considered divine, then one would never allow oneself to analyze it, dissect it, or perform experiments on it, and thus all incentives for doing science would be suppressed. But if the world is indeed God's creation, then we can "interrogate" the world by investigation, exploration, and experiment.

Seen in this light, it can hardly be called a coincidence that belief in a Judeo-Christian God made science actually possible. Faith in this one God changes the universe—once seen as inhabited by various spirits, deities, and goddesses—into something "rational," open to further exploration. The book of Wisdom says about God, "Thou hast arranged all things by measure and number and weight."[10] Hence the only way to find out what the Creator has actually done is to go out, look, and measure—which are necessary conditions for scientific exploration and experimentation.

The world is orderly

Not only is science a very orderly enterprise in itself, but it also reveals a very orderly universe. Interestingly enough, the Greek word "*cosmos*"—a synonym for the Latin word "*universum*"—means "order." The role of hypotheses, theories, and laws of nature in science is to capture this very order, which otherwise might easily elude us. Thanks to science, we see a striking order in the world around us.

Examples are plentiful. Planets move around the sun in very orderly orbits, with all these orbits lying in almost the same plane, making the solar system look like a giant platter. Or think of litmus paper, which we use to test the pH of liquids. Red litmus paper turns blue when the pH is alkaline, while blue paper turns red when the pH is acidic. They do so always and everywhere. Even something as simple as a snowflake shows a very intricate and orderly pattern. We could add numerous other examples. They all tell us there are definitely order and regularity in this world. If we find exceptions

[10] Wisd. 11:20.

to this order, we have to find out how to explain the exceptions. In other words, exceptions don't make the world disorderly, but they force us to come up with a more accurate order in our explanations.

Science reveals something very astonishing about this world: namely, that it's not a chaotic entity. Einstein was right when he wrote in a letter dated March 30, 1952, "A *priori*, one should expect a chaotic world."[11] But instead of expecting chaos, scientists assume that the world has a remarkable order. The laws of nature that science brings to light are the many patterns of regularity and harmony we have detected in this world. One such regularity, discovered by Galileo, is that two rocks, dropped at the same time from the same height, reach the ground at the same time—not only in Pisa in Galileo's time, but everywhere on earth. The assumption that these regularities exist makes science possible. No wonder Einstein said in one of his interviews about his formula $E = mc^2$, "If I hadn't an absolute faith in the harmony of creation, I wouldn't have tried for thirty years to express it in a mathematical formula."[12]

Yet it seems to be tempting to think that order comes out of chaos. Those who think so may refer to the emergence of the universe from the "chaos" of the Big Bang. Or they may point to what happens when the temperature of water is lowered to its freezing point: the chaotically swirling water molecules begin to line up in a striking pattern. It looks as if the orderly pattern of ice is coming out of the chaotic movement of molecules.

Nevertheless, the truth is that chaos can never create the order found in this world, just as blindness can never create sight. The popular saying "Garbage in, garbage out" applies even here. The

[11] Albert Einstein, *Letters to Solovine*, trans. Wade Baskin (New York: Philosophical Library, 1987), 132–133.

[12] William Hermanns, *Einstein and the Poet: In Search of the Cosmic Man* (Wellesley, MA: Branden Books, 2013), 68.

order we see in nature does not and cannot come from chaos, but must come from a more fundamental, preexisting order at a deeper level. Order can only come from order.

In other words, order has to be built in for order to come out. Shaking jars of variously shaped candies won't create much more order, for there is no order in their shapes to begin with. But when shaking a jar with round candies, we do create more order, as round candies have an underlying order—that is, at a "deeper" level"—which allows for a so-called "hexagonal closest packing" structure. The underlying order of round candies explains the order generated by our shaking the candies.[13]

In this example, physics requires the round candies to lower their gravitational potential energy as much as possible by creating the geometry of hexagonal packing. It is this underlying order that enables and explains the order of round candies' forming the closest packing structure. In other words, the preexisting order inherent in the round candies is greater than the order that emerges after the candies arrange themselves. It is not an order imposed by scientists but an order waiting to be unveiled by them. And it is certainly not an order coming out of chaos.

As a matter of fact, one cannot even comprehend the word "chaos" without, at the same time, contrasting it with the word "order"—chaos is that which lacks order. We speak of "chaos" when we haven't yet been able to understand or find out the order behind it. When something seems chaotic—for instance, the transmission of genetic traits to the next generation—science will try to find an orderly, systematic explanation. It is science's task to unearth the order behind our observations. There is no way scientists can say

[13] This analogy originated with particle physicist Stephen Barr. Stephen M. Barr, "Fearful Symmetries," *First Things* (October 2010): 26.

that, after a long search, they have discovered there is actually no order behind what they are investigating. If they do say so, they should be told to keep searching for the order that's apparently still eluding them.

As a matter of fact, the order of the universe is a fundamental assumption of science. It is only due to the orderly design of the universe that scientists make explanations and predictions—which would be impossible in a chaotic, irregular world. In other words, order must come first, before science can even get started. Order is not something scientists discovered after increasingly successful searches revealed more and more cases of order. There is no way we can prove order by adding more and more confirming cases to the collection.

Order cannot even be falsified by cases where no order has been found (yet). In other words, order is not a scientific discovery, but rather a philosophical assumption that enables science and propels scientific research. What may seem chaotic to laypeople will turn out to be very orderly as science progresses. Those who reject any order in the universe have basically given up on science. If there were no order in the universe, it would make no sense to search for laws of nature in physics, chemistry, biology, and other disciplines.

What seems to belie all of this is the fact that many scientists recently developed a strong interest in chaos and chaotic systems, as if these could falsify the existence of "law and order" in the universe. They point out that some natural systems can only be described by non-linear mathematical equations with such complex solutions that we cannot predict exactly what the system will do in the near future. Or, to take another example, our measurements of all the initial conditions of a particular system (for example, hurricanes or tornadoes) may be too numerous or too inaccurate to predict what exactly the outcome will be.

However, this is not really chaos; it only appears to be. In fact, these scientists are still looking for the very order behind seemingly chaotic phenomena. When the weather forecast is off the mark, we do not conclude that the weather is unpredictable or shows unorderly behavior. We just do not know enough to be perfectly accurate in our predictions. But that's not chaos!

The question is again: How does the assumption of order point to God? The legendary astronomer Johannes Kepler couldn't have said it better: "The chief aim of all investigations of the external world should be to discover the rational order and harmony which has been imposed on it by God."[14] Order points to an orderly Creator: God.

There is indeed an amazing harmony between the order we find in the world and the order we assume in our scientific knowledge of the world—that is, between the physical order in the world and the rational order in our minds. Somehow, the human mind is such that it has access to the world and can grasp reality the way it is—orderly. Even the equations scientists use in physics, chemistry, and biology reflect how orderly the world is. These equations don't impose, but reflect order. Consequently, the harmony between the rational order found in the human mind and the physical order found in the world is so striking that this strongly suggests they both have the same origin and the same Author, which in turn strongly points to God.

As the physicist Paul Davies once said, "There must be an unchanging rational ground in which the logical, orderly nature of the universe is rooted."[15] By using the word "must," he suggests

[14] Johannes Kepler, *De Fundamentis Astrologiae Certioribus*, thesis XX.

[15] Paul Davies, "What Happened before the Big Bang?," in *God for the 21st Century*, ed. Russell Stannard (Philadelphia: Templeton Foundation Press, 2000), 10–12.

this conclusion is hard (albeit not impossible) to deny. Davies is basically asking the following rhetorical question: Could there be order in this world without an no orderly Creator? His answer is no.

How can that be? When we speak of order, we intuitively think of someone who imposed the order, for we could easily imagine a chaotic world without any order. As a matter of fact, the world we live in doesn't have to behave the way it does, yet we do find a very specific "law and order" regulating what happens and can happen in this universe. That's something hard to explain, unless the order of the world does point to an orderly Creator, God. Without the order of creation, without the implementation of laws of nature, the world could easily be a world of real chaos.

Pope John Paul II could not have summarized this better: "Those who engage in scientific and technological research admit, as the premise of its progress, that the world is not a chaos but a cosmos; that is to say, that there exist order and natural laws which can be grasped and examined."[16] Obviously, the pontiff was pointing to God as the source of all of this.

The world is comprehensible

Albert Einstein used to say that the most incomprehensible thing about the universe is that it is comprehensible; he went so far as to speak of a mystery.[17] It is indeed incomprehensible that we can, at least in principle, comprehend the universe the way it is. The world could just as well be a complete enigma to us, but somehow—quite fortunately, I would say—it's not.

[16] John Paul II, Discourse to the Pontifical Academy of Sciences, October 31, 1992, in *Papal Addresses*, 343.

[17] Albert Einstein, *Out of My Later Years* (1950; repr., New York: Gramercy, 1993), 64.

The idea that this universe is comprehensible, or intelligible, certainly does not come from science itself. Scientists assume that the world can be understood and taken as intelligible—otherwise there would be no reason for them to pursue science and make the world intelligible. They don't know yet completely what this intelligible world looks like, but they do know—that is, they assume—that the world must be intelligible in principle. So the idea of a comprehensible world is definitely not the outcome of intense and extensive scientific research; it is, again, an assumption that must come first, before science can even begin. It does not have to be confirmed over and over, but it is a precondition for confirmation to work.

This assumption is so basic to science that it easily eludes scientists. If you were told that some scientists had discovered that certain physical phenomena are not intelligible, you would, or at least should, tell them to keep searching and come up with a better hypothesis or theory—based on this fundamental philosophical knowledge that says the universe is essentially intelligible and comprehensible. Science calls for this assumption. Who would want to start studying a phenomenon that we could never understand at all? It is only because of their trust that nature is comprehensible in principle that scientists have reason to trust their own research.

Developments in quantum physics seem to belie this assumption. In quantum theory, the basic units of light are photons or light quanta, which are considered to be both of a wave nature and a particle nature. This property is known as wave-particle duality. In an absolute sense, then, light is actually neither a particle nor a wave, but only exhibits wave or particle features. That seems to defy comprehensibility. How can waves have particle-like characteristics and particles have wave-like characteristics? That's counterintuitive. However, that didn't stop scientists from making the wave-particle duality comprehensible. The physicist

Paul Dirac, for example, speaks of "the wave function giving us information about the probability of our finding the photon in any particular place when we make an observation of where it is."[18] The quest for comprehensibility prevents science from becoming a mere riddle.

How, then, do scientists make this world more intelligible for us? By developing more and more complex *theories*. They have come up with theories about atoms and molecules, about electrons, neurons, and genes, to name just a few. But they have also formulated more and more complex *equations*.

Theories and equations, however, do not necessarily make reality more intelligible. Let me explain. Because scientists, especially physicists, are "specialists" in measuring, quantifying, and counting, they tend to express their theories with mathematical equations and formulas. Although physics both starts and ends in the physical reality, it submits the measurements it has drawn from the physical world to the rules of mathematics—that is, for the most part, it attempts to explain physical things insofar as they can be described mathematically in equations. This emphasis on quantification has contributed to the "mathematicalization" of science, wherein everything is reduced to mathematical equations and formulas.

This process began with Italian astronomer and physicist Galileo Galilei. He famously said that the Book of Nature "is written in the language of mathematics."[19] This belief in the power of mathematics has only become stronger and stronger ever since. Paul Davies puts it this way: "Scientists do not use mathematics merely as a convenient way of organizing the data. They believe

[18] Paul Dirac, *Principles of Quantum Mechanics*, 4th ed. (Oxford: Clarendon Press, 1982), 9.

[19] Galileo, *Il saggiatore* (The Assayer), trans. Thomas Salusbury, 178.

that mathematical relationships reflect real aspects of the physical world."[20]

It's actually amazing that our world behaves in accordance with our equations. The laws of gravitation and electromagnetism, the laws that regulate the world within the atom, the laws of motion—all are expressed as tidy, though sometimes complex, mathematical relationships. They make the physical world intelligible for us in mathematical terms. With these equations, scientists are able to explain and predict what is happening in the real world. Somehow, equations reveal to us an elegant mathematical order in the world around us. As physicists probe to a deeper level of subatomic structure, they expect to encounter more and more elegant mathematical order.

No doubt, sometimes we understand things better when they can be mathematically expressed in equations—at least, physicists do: mathematics makes science more precise and testable. But the downside of this increasing mathematicalization is that these equations leave much of reality behind. It is not always the case that mathematics makes things more intelligible and comprehensible. Bertrand Russell once said, "Physics is mathematical not because we know so much about the physical world, but because we know so little."[21]

Physics seems to have become a matter of manipulating symbols in mathematical equations. Since then, we have come to know the world through equations and, thus, have lost direct contact with reality as we know it. Michael Augros mentions his experience of how his physics teacher in high school explained the difference

[20] Paul Davies, "Is nature mathematical?," *New Scientist*, March 21, 1992, 25–27.

[21] Bertrand Russell, "The Nature of Our Knowledge of Physics," chap. 15 in *An Outline of Philosophy*, Routledge Classics edition (New York: Routledge, 2009).

between *mass* and *weight* with the question, "Would you rather lift my car or push it?"[22] That's how he learned that mass resists your push, and weight resists your lift. That step in understanding seems to be necessary before the equations can do their work of explicating reality.

Bertrand Russell said it well: "All that physics gives us is certain equations giving abstract properties of their changes. But as to what it is that changes, and what it changes from and to—as to this, physics is silent."[23] Now, if physics gives us only the mathematical structure of material reality, then not only does it not tell us everything there is to know about material reality, but it implies that there must be more to material reality than what physics tells us.

And again, the question is this: How does the assumption of intelligibility point to God? When speaking of "assumptions" in science, we might be creating the impression that they exist only in the minds of scientists—as if these scientists decided to assume certain things in order to legitimize their research. In the eyes of scientists, each assumption underlying science does indeed look like an assumption that they happen to believe in. But if that's all there is to it, science becomes a pretty shaky enterprise, resting on man-made quicksand. Only God, one can well argue, can give these assumptions a firm basis. Without a Creator God, scientists would fundamentally lose their reason for trusting their own scientific reasoning.

This would be no surprise to religious believers, but even scientists have increasingly come to the conclusion that the universe has an elegant, intelligible, and discoverable underlying physical

[22] Michael Augros, *Who Designed the Designer? A Rediscovered Path to God's Existence* (San Francisco: Ignatius Press, 2015), 28.

[23] Bertrand Russell, *My Philosophical Development* (London: Routledge, 1959), 13.

and mathematical structure. The beauty and elegance of the laws of nature and the mathematical equations behind them point to a Divine Intellect who created them. Even Steven Weinberg, a theoretical physicist, Nobel laureate, and self-declared atheist, had to admit that "sometimes nature seems more beautiful than strictly necessary."[24] He also said, "Mathematical structures that confessedly are developed by mathematicians because they seek a sort of beauty are often found later to be extraordinarily valuable by the physicist."[25] That in itself is amazing.

The fact that the universe has an elegant, intelligible, and discoverable underlying mathematical and physical structure calls for some kind of explanation—or otherwise must be left unexplained. No wonder, then, that Albert Einstein had to acknowledge, "Everyone who is seriously involved in the pursuit of science becomes convinced that some spirit is manifest in the laws of the universe, one that is vastly superior to that of man."[26] The late astrophysicist Sir James Jeans put it this way: "The universe begins to look more like a great thought than a great machine."[27] And Stephen Hawking famously noted that all theoretical physics can do is give us a set of rules and equations that correctly describe the universe; it cannot tell us *why* there is any universe for those equations to describe.[28]

[24] Steven Weinberg, *Dreams of a Final Theory* (New York: Pantheon Books, 1992), 250.

[25] Ibid., 153.

[26] *Dear Professor Einstein: Albert Einstein's Letters to and from Children* (Amherst, NY: Prometheus Books, 2002), 129. Also quoted in Max Jammer, *Einstein and Religion: Physics and Theology* (Princeton, NJ: Princeton University Press, 1999), 93.

[27] James Jeans, *The Mysterious Universe* (Cambridge, UK: Cambridge University Press, 1930), chap. 5.

[28] Stephen Hawking, *A Brief History of Time* (New York: Bantam Books, 1988), 174.

The fact that the theories and equations scientists have discovered form a single magnificent edifice of great subtlety, harmony, and beauty makes Stephen Barr declare, "The question of a cosmic designer seems no longer irrelevant, but inescapable."[29] In the same vein, the British physicist John Polkinghorne speaks of "pointers to the divine as the only totally adequate ground of intelligibility."[30]

What remains baffling about all of this is that our world behaves in accordance with our scientific theories and equations. They came from the rational minds of scientists and yet apply to what happens in the physical world. Somehow, there is reason in the human intellect, and there is reason in the world around us. How come they match each other so beautifully? The only rational answer is this: they both derive from the Reason we find in the Divine Intellect. That's exactly the answer religion gives us: we were created in God's image and likeness.[31] Only this can explain the power of reason in both the physical world and the human intellect. If God is indeed the Creator of this world, then God's rationality can be found in His creation, just as it can be found in the rationality of the human mind. No wonder, then, that there is a correspondence between the two.

Pope John Paul II's encyclical *Fides et Ratio* offers a similar view as to how the activity of science is possible: "It is the one and the same God who establishes and guarantees the intelligibility and reasonableness of the natural order of things upon which scientists confidently depend, and who reveals himself as the Father of our Lord Jesus Christ."[32]

[29] Stephen Barr, "Retelling the Story of Science," *First Things* 131 (March 2003): 16–25.

[30] John Polkinghorne, *Science and Creation: The Search for Understanding* (West Conshohocken, PA: Templeton Press, 2006), 19.

[31] Gen. 1:26a.

[32] Pope John Paul II, encyclical letter *Fides et Ratio* (Faith and Reason) (September 14, 1998), no. 34.

The world is more than we know

Human knowledge is limited. If the world is no more than what we know, then we have no reason to search any further—our knowledge and our search would have come to an end. But they haven't—and never will. There is always more to know and more to search for.

In other words, science is always a work in progress. Nevertheless, some scientists cannot resist the temptation to claim certainty and finality. The late Dutch physicist Pieter Zeeman, who discovered the "Zeeman effect"[33] and later became a Nobel laureate, was fond of telling how in 1883, when he had to choose what to study, people had strongly dissuaded him from studying physics. "That subject's finished," he was told. "There's no more to discover."[34]

It is even more ironic that this also happened to Max Planck, since it was he who, in 1900, laid the foundations for one of the greatest leaps in physics, the quantum revolution.[35] Apparently, it remains a timeless temptation to claim that the unknown has been reduced to almost nothing. However, the magnitude of the unknown is, well ... unknown! Science is by its nature a work in progress.

The late American physicist John A. Wheeler put it this way: "We live on an island surrounded by a sea of ignorance. As our island of knowledge grows, so does the shore of our ignorance."[36]

[33] The effect by which a spectral line splits into several components in the presence of a static magnetic field.

[34] Mentioned by the physicist Anthony Van den Beukel, *The Physicists and God* (North Andover, MA: Genesis Publishing, 1996), 37.

[35] From a 1924 lecture by Max Planck, printed in *Scientific American* 274, no. 2 (February 1996): 10. See Alan P. Lightman, *The Discoveries: Great Breakthroughs in 20th Century Science, Including the Original Papers* (Toronto: Alfred A. Knopf, 2005), 8.

[36] Quoted in Clifford A. Pickover, *Wonders of Numbers* (New York: Oxford University Press, 2001), 195.

This means that those who think they know everything and have answers for everything must live on one tiny island.

As it turns out, there is always more for us to know. We need to realize there are at least two kinds of limitations as to what science can discover. First, there is the fact that the magnitude of the unknown in science is necessarily unknown, as we found out. But there is also a second, and more important, limitation for science. It is the fact that science does not and cannot have answers to *all* our questions.

Nevertheless, some scientists have made up another assumption—but this time an unwarranted one—that the scientific method is not only the best method there is, but also the only method we have to understand the world. They claim that "the real world" is only a world of quantified material entities. This leads them to believe that all our questions have a scientific answer phrased in terms of particles, quantities, and equations. Their claim is that there is no other point of view than the "scientific" worldview. They believe there is no corner of the universe, no dimension of reality, no feature of human existence beyond its reach. In other words, they have a dogmatic, unshakable belief in the omni-competence of science.

Could that really be true? Let's not forget, as it has been famously said, that "gravitation is not responsible for people falling in love."[37] The Brazilian-born particle physicist Marcelo Gleiser, who earned the Templeton Prize in 2019, put it this way: "What should change is a sense of scientific triumphalism—the belief that no question is beyond the reach of scientific discourse."[38]

[37] This quote is usually attributed to Isaac Newton, but also to Albert Einstein (jotted in German on the margins of a letter to him) as cited in Helen Dukas and Banesh Hoffmann, eds., *Einstein: The Human Side* (Princeton, NJ: Princeton University Press, 1981), 56.

[38] Marcelo Gleiser, "How Much Can We Know?" *Scientific American* 318, no. 6 (June 2018): 72–73.

Because there is more to the world than what science can discover, we need to counteract scientific triumphalism. The late UC Berkeley philosopher of science Paul Feyerabend, for instance, was doing exactly that when he said, "Science should be taught as one view among many and not as the one and only road to truth and reality."[39] Even the "positivistic" philosopher Gilbert Ryle expressed a similar view: "The nuclear physicist, the theologian, the historian, the lyric poet and the man in the street produce very different, yet compatible and even complementary pictures of one and the same 'world.'"[40] Science provides only one of these views.

Interestingly, the astonishing successes of science have not been gained by answering every kind of question, but precisely by refusing to do so. One could even agree with the late biologist and Nobel laureate Konrad Lorenz that a scientist "knows more and more about less and less and finally knows everything about nothing."[41]

How does the existence of scientific limits point to God's existence? Once we realize the limitations of our scientific knowledge, we "know" there are things we don't know, and we "know" there is more to know. "Limitations" mean that we have the capacity to look beyond and step outside the territory of science. Apparently, we can transcend what we know. God's existence, for instance, is something we can find only beyond the horizon of our knowledge.

The pioneer of quantum physics, Max Planck, said something similar: "Science cannot solve the ultimate mystery of nature. And that is because, in the last analysis, we ourselves are a part of the

[39] Paul Feyerabend, *Against Method: Outline of an Anarchistic Theory of Knowledge* (New York: Verso Books, 1975), viii.

[40] Gilbert Ryle, *Dilemmas* (Cambridge, UK: Cambridge University Press, 1960), 68–81.

[41] Larry Collins and Thomas Schneid, *Physical Hazards of the Workplace* (Boca Raton, FL: CRC Press, 2001), 107.

mystery that we are trying to solve."[42] A mystery is not something about which we can't know anything, but something about which we can't know everything. As Stephen Barr explains, a mystery is something that "cannot be seen, not because there is a barrier across our field of vision, but because the horizon is so far away. It is a statement not of limits, but of limitlessness."[43]

Somehow, the limitations of our scientific knowledge point to something or someone that does not have such limitations. We obviously have the capacity to transcend who we are and what we know. Think of this: when humans say "I am only human," they are not comparing themselves with something "below" them, such as cats, dogs, or apes, but they are comparing themselves with what is "above" them and transcending them. When I call myself "only human," I am actually comparing myself with a Person who does not have the limitations I experience myself. In some mysterious way, I am reaching out into the realm of the Absolute, far beyond myself. In doing so, the "finite" catches a glimpse of the Infinite.

Something similar happens when we approach the limits of science. We become aware of what's outside those limits. Limits point to what is unlimited—God. One of those limits is that science itself can never prove God's existence for the simple reason that science is about material things, whereas God is not a material being. However, like most human endeavors, science uses certain assumptions, which are not material, yet are needed for science to do its work. Perhaps those assumptions point somehow to God's existence. Although they cannot prove God's existence, they are pointers to Him.

[42] Quoted in Ken Wilber, ed., *Quantum Questions* (Boston: New Science Library, 1984), 153.

[43] Stephen Barr, *Modern Physics and Ancient Faith* (Notre Dame, IN: University of Notre Dame Press, 2003), 14.

It seems to me that the physicist Marcelo Gleiser was hinting at this when he said:

> If you look carefully at the way science works, you'll see that yes, it is wonderful—magnificent!—but it has limits. And we have to understand and respect those limits. And by doing that, by understanding how science advances, science really becomes a deeply spiritual conversation with the mysterious, about all the things we don't know.[44]

So we must come to the conclusion that science cannot even begin without certain assumptions. What science cannot do is prove its own assumptions, yet they are needed for science to do what it can do. In other words, what science can do is possible only if it realizes what it cannot do—test and prove its own assumptions, to begin with.

[44] Marcelo Gleiser, "Atheism Is Inconsistent with the Scientific Method, Prizewinning Physicist Says," interview by Lee Billings, *Scientific American*, March 20, 2019.

2

How Laws of Nature Point to God

In the previous chapter, we mentioned a few times the so-called "laws of nature." It is a common expression among scientists, yet the term "law" has become controversial in this context. Let's find out what scientists mean when they call something a "law of nature," and why there is some controversy about the issue.

A world of cause and effect

The simplest kinds of experiments in science study the relationship between two variables, usually expressed in terms of cause and effect. Take, for instance, Boyle's law, which describes how the pressure of a gas tends to increase as the volume of the container decreases. So gas pressure is related to its volume. Specifically, they are inversely proportional to each other. These are two variables with a cause-and-effect relationship.

When the relationship between two variables is tested during an experiment, scientists often speak of an independent variable, which is the one freely chosen by the scientist in an experiment, and a dependent variable, which is the one we study so as to see how it is impacted by the other. We could call the independent variable the cause of the effect that we observe in the dependent one.

This relationship can easily be generalized in the following statement: everything has a cause in such a way that like causes have like effects. This is a principle that has basically ruled science from the very beginning. It's a principle that doesn't need to be tested over and over again. Scientists "know" that it makes experiments possible. The principle of "like causes' having like effects" would read in translation as follows: from causes that appear similar, scientists *expect* similar effects; if there are effects, then there must be certain causes involved; and if the effects are different, then they must have had different causes. Thanks to the principle of "like causes' having like effects" scientists can experiment, explain, and predict.

The general conviction behind this principle is that we live in a world where *causality* reigns—nothing comes from nothing. This assumption is obviously order-related, but there is more to it. You don't have to be a scientist to understand that like causes have like effects. All of us seem to know this rule almost intuitively, as when we heat the water on the stove, for instance. But scientists in particular make it their profession to apply this rule methodically. In their research, they attempt to unveil the patterns of causality in the world around us.

But there is a potential problem here: we do not *see* causality—similar to the way we do not see gravity or electricity or black holes. It's true, we do not see causation in the same way in which we see colors and shapes and motion. So what is causality, then? Could it be something we imagine, something only in our minds?

The eighteenth-century skeptical philosopher David Hume—mind you, not a scientist—went that road. All talk about cause-and-effect relationships became discredited when Hume came along. Since the supposed influence of a cause upon its effect is not directly evident to our senses, Hume concluded that the connection between cause and effect is not an aspect of the real world, but only a habit of our thinking as we become accustomed to seeing one thing constantly conjoined to another. Thus, Hume reduced causality to correlation at best—no longer something real found in the world outside ourselves, but rather a way of thinking about the world.

Does Hume have a point here? His view has certainly not been embraced by great physicists such as Max Planck and Albert Einstein, who all assume that physics describes a reality independent of ourselves, and that the theories of science show not only how nature behaves but why it behaves exactly as it does and not otherwise. And almost all other scientists do not side with Hume.

Besides, Hume's analysis would erase the important distinction between causation and correlation by reducing both to a series of mere subjective associations or generalizations. One of the first things you learn in a science class is that correlation doesn't imply causation. Scientists always need to make sure that causality is not just a matter of correlation. Just because two variables or factors are correlated does not necessarily mean that one causes the other. Correlation may be useful for prediction, but it does not always give us an explanation of a causal connection, if there is any. Besides, correlation is symmetrical, but causation only goes one way.

A well-known case is the correlation between smoking tobacco and the occurrence of lung cancer. In order to call tobacco a mutagen or carcinogen, scientists had to prove causation, not just correlation. That's easier said than done. It could well be that the increased use of tobacco and the increased use of nylon stockings, for instance, are equally associated with lung cancer. In addition, there could be many other causal factors involved: the increased rate of lung cancer could have been the result of better diagnosis or increased industrial or car-exhaust pollution;

perhaps people who were more genetically predisposed to smoking were also more susceptible to getting cancer. It took a large study involving more than forty thousand doctors in the UK to show conclusively that smoking really does cause cancer and is not just a matter of correlation.

So where did Hume go wrong, then? When we see the sun rise every morning, for instance, we know that there is not a different sun rising every morning. Although we know the world through sensations (or sense impressions, à la Hume), the truth of the matter is that they are just the media that give us access to reality. A current country fellow of Hume, the philosopher John Haldane, put it well when he said, "One only knows about cats and dogs through sensations, but they are not themselves sensations, any more than the players in a televised football game are color patterns on a flat screen."[45] Hume actually came to his view because he took causality as rooted not in the identity of acting *things*, but in a relationship between *events*, assuming that "all events seem entirely loose and separate. One event follows another; but we never can observe any tie between them. They seem *conjoined*, but never *connected*."[46]

It should be argued instead that the actions an entity can take are determined by what that entity *is*. When one billiard ball strikes another, it sends the other rolling because of the nature of the two balls and their surroundings. This entails the following: when we know that billiard balls are solid, and when we see one ball moving toward another, then certain effects are quite impossible. The moving ball cannot, for example, just pass through the second

[45] John Haldane, "Hume's Destructive Genius," *First Things* 218, (December 2011): 23–25.

[46] David Hume, "Of the Idea of necessary Connexion," part II in *An Enquiry Concerning Human Understanding*, Harvard Classics, vol. 37, no. 3 (New York: P.F. Collier and Son, 1909–1914).

ball and come out the other side continuing at the same speed; nor can the first ball stop at exactly the same place as the second ball; nor can one of the balls suddenly vanish, and so on and so forth. The qualities of the balls determine the kind of effect that the impulse of the first ball as a cause will have on the second. In other words, when we see entities acting, we see causality in terms of "like causes' having like effects."

It is simply very hard, if not impossible, to get rid of cause-and-effect relationships. They describe real patterns of causality in the real world. In other words, causality is not something we fancy in our minds, but something we see operating in the real world. Anyone coming up with an exception to the rule of causality needs to search further and better, rather than abandoning the principle of causality. The general statement "like causes have like effects" admits no exceptions, because we "know" that nothing can happen without a cause.

One could even argue that it is logically impossible to prove that something has no cause at all, since searches like these never reveal the *absence* of their object. Causality can never be conclusively defeated by experiments since causality is the very foundation of all experiments. Science can never scientifically prove that there is causality in this universe, so it must accept that there is. Science cannot get started without the principle of "like causes' having like effects."

Yet the idea that nothing happens without a cause has been questioned recently by new developments in quantum physics. Quantum theory seems to suggest the possibility that some phenomena have no cause. The so-called "uncertainty principle" of quantum mechanics, for instance, says that an object in the quantum world cannot have both a well-defined position and a well-defined speed: the more accurately one measures the position, the less accurately one can measure the speed, and vice versa.

Let me mention first that there are at least seventeen different interpretations of quantum theory, so we shouldn't put all our stock in one of them.[47] These are claims *about* quantum theory rather than claims *of* quantum theory. Probably the most popular interpretation comes from the Danish physicist Niels Bohr, who went as far as claiming that physical systems generally do not have definite properties prior to being measured. Bohr is basically saying that reality does not exist when we are not observing it.[48]

The problem is, though, that our perceptual experiences give us knowledge of the external physical world only because they are *causally* related to that world. To deny causality in the name of science would therefore undermine the very empirical foundations of science. How could we ever account for our knowledge of the world that physics describes if we have no causal contact with the world at all? Albert Einstein, for one, always resented and resisted this implication.[49]

But even if we do accept this interpretation and deny the existence of causes in the quantum realm, we need to realize first of all that quantum events may not have a *deterministic* cause, but this doesn't imply they have no cause at all. Second, even quantum events still have an explanation based on the laws of quantum mechanics—it is these laws that make quantum phenomena intelligible. Third, even if certain events do not have a completely determinative *physical* cause, that does not mean, according to the particle physicist Stephen Barr, that these events have no cause

[47] See, for instance, Graham P. Collins, "The Many Interpretations of Quantum Mechanics," *Scientific American*, November 19, 2007.

[48] Let's call this the "idealist" interpretation, which was promoted by Niels Bohr and Werner Heisenberg, among others.

[49] Let's call this the "realist" interpretation, represented by physicists such as David Bohm, John Bell, and Richard Feynman.

whatsoever, for not all causes have to be physical causes.[50] So there seems to be no reason to abandon the causality principle, not even in quantum physics.

And again, the question is this: How does the existence of causality in our world point to God? Causality is closely connected with the assumption of order and regularity in the universe. If like causes do have like effects, there must be something in nature that makes this happen. There is something very real about cause-and-effect relationships. What is it, then, that makes them real? There must be an ultimate cause that has the effect of making all cause-and-effect relationships real. This ultimate cause has traditionally been called the First Cause, which refers to God.

This insight is probably best known as one of St. Thomas Aquinas's proofs of God's existence.[51] In this proof, St. Thomas is thinking in terms of cause and effect. He comes to the conclusion that it is the First Cause that causes the existence of secondary causes in this world and makes them possible. Without a First Cause, none of the secondary causes could exist, let alone become causes of their own.[52] Aquinas refers to this First Cause as God: "God is in all things the cause of being."[53] Without God, all (secondary) causes wouldn't be and could not exist. So the principle of causality itself unmistakably points to God, the First Cause.

[50] Stephen Barr, *Modern Physics*, 264.

[51] It is deceivingly called the "argument from motion." When Aquinas says that A "moves" B, he is actually saying that A "causes" B and thus "explains" B. There is no movement through space here.

[52] I explained this much more extensively and rigorously in my previous book: Gerard Verschuuren, *A Catholic Scientist Proves God Exists* (Manchester, NH: Sophia Institute Press, 2020).

[53] Thomas Aquinas, *Summa Contra Gentiles* II, 46.

Are the laws of nature really laws?

The use of the term "law" is quite common among scientists. They speak of the law of gravitation, the three laws of motion, the ideal gas laws, even Mendel's laws.

In general, it could be said that laws of nature typically express a relationship between two or more variables. However, this relationship can be either deterministic or probabilistic. Deterministic laws of nature predict one specific effect or outcome. A few examples are the law of Archimedes,[54] Newton's laws, Ohm's law,[55] Boyle's law,[56] and Joule's first law.[57] In these laws, the outcome is specifically determined.

There are also probabilistic laws, which state that, on average, a certain fraction of cases displaying a given condition will display a certain other condition as well. These laws concern the probabilities of certain effects. Examples are the law of radioactive decay,[58] Mendel's law of segregation,[59] and the law of natural selection,[60]

[54] It states that the upward buoyant force that is exerted on a body immersed in a fluid is equal to the weight of the fluid that the body displaces and acts in the upward direction at the center of mass of the displaced fluid.

[55] It states that the current through a conductor between two points is directly proportional to the voltage across them.

[56] It states that the pressure of a gas increases as the volume of its container decreases.

[57] It states that the power of heat generated by a conductor is proportional to the product of its resistance and the square of the current.

[58] It states that the probability per unit time that an unstable nucleus will decay is a constant; it is independent of time.

[59] It states that the two versions of each gene are segregated from each other on a random basis, so that only one from each parent goes to the next generation.

[60] It states that differences in genetic features cause differential chances of reproduction and survival.

or a quantum law that expresses the probability that a particle will be found at a certain location.

Such laws are an essential part of science—there is no science without them. "The concept of law," Paul Davies says, "is so well established in science that until recently few scientists stopped to think about the nature and origin of these laws; they were happy to simply accept them as 'given.'"[61]

But times are changing. Many scientists do not really think of them as "laws" anymore. They seem to have developed some form of law phobia. At one point, it was very common to search for laws of nature. Johannes Kepler, one of these old-timers, said, "Those laws [of nature] are within the grasp of the human mind; God wanted us to recognize them by creating us after his own image so that we could share in his own thoughts."[62] But that was quite a while ago. Nowadays, the term "law" has become contaminated. But by what?

Probably the first reason for this contamination is the apparent connotations of the term "law." The term reminds some scientists too much of laws implemented by a higher authority, an external lawgiver. Many prefer not to use the expression "laws of nature," because they dislike what it suggests—a Lawmaker from above. No wonder, then, that scientists who want to keep God out of science dislike, or at least avoid, the term "laws of nature." They think religion has infiltrated science too much. It's obviously ideology, not science, that makes them believe this.

[61] Paul Davies, *The Mind of God: Science and the Search for Ultimate Meaning* (New York: Simon and Schuster, 1992), 73.

[62] Johannes Kepler, "Letter of 9/10 Apr 1599 to the Bavarian chancellor Herwart von Hohenburg," in *Johannes Kepler: Life and Letters*, trans. and ed. Carola Baumgardt (New York: Philosophical Library,1951), 50.

True, religion did play quite a role in the early development of science. The first scientists were religious believers who connected their scientific work with God's work in nature. Max Planck, who revolutionized physics with his quantum theory, was right when he said that "the greatest naturalists of all times, men like Kepler, Newton, Leibniz, were inspired by profound religiosity."[63] Copernicus's revolutionary discovery of the heliocentric model, for instance, was in fact based on his religious belief that nothing was easier for God than to have the earth move, if He so wished: "To know the mighty works of God, to comprehend His wisdom and majesty and power ... surely all this must be a pleasing and acceptable mode of worship to the Most High, to whom ignorance cannot be more grateful than knowledge."[64] And Kepler's Christian belief, too, told him God would not tolerate the inaccuracy of circular models of planetary movements in astronomy, so he replaced circular orbits with elliptical ones[65]—which made him exclaim, "Through my effort God is being celebrated in astronomy."[66]

A second reason for dismissing the term "law" in science is the damage David Hume caused. He had seriously tried to get rid of laws of nature by claiming that they were just creations of the mind. If Hume were right, laws of nature would exist only in the minds of physicists, chemists, and biologists—not in the world itself. This solution fails to explain the fact that laws actually do hold true in the real world. How is it possible for a bridge that has been designed according to the right laws to stand firm, whereas another

[63] Max Planck, *Religion und Example* (Leipzig: Johann Ambrosius Barth Verlag,1937), 332.

[64] Quoted in Francis Collins, *The Language of God: A Scientist Presents Evidence for Belief* (New York: Free Press, 2006), 230–231.

[65] Johannes Kepler, *Astronomia Nova* (New Astronomy), chap. 19, 113–114.

[66] Letter to Michael Maestlin, dated October 3, 1595.

bridge collapses because its engineers erred in their calculations or used the wrong laws? Competent engineers may and probably do have better mental habits than inept ones, but that is not the only difference between them or even the most important. These laws could never hold if they were only creations of the human mind. That is the reason why laws of nature have to be discovered, over and above being invented in the mind.

A third explanation for the demise of the term "law" can be found in the attempts of some scientists and philosophers to interpret the so-called laws as mere *descriptions* of certain regularities in the universe. They can be expressed in equations that are no longer connected with reality. In this view, Boyle's law, for instance, merely describes—in the form of either words or equations—how the pressure of a gas is associated with the volume of its container. In this interpretation, there is no longer anything enigmatic in scientific laws, for they just describe cause-and-effect relationships without any further implications. They have basically become a case of correlation instead of causation.

However, we discussed already that correlation is very different from causation. Besides, even if we declare the laws of nature to be pure descriptions of cause-and-effect relationships, we are still left with the question as to where this regularity comes from. Regularity cannot explain itself. John Stuart Mill had said already in 1865 that "the laws of Nature cannot account for their own origin."[67] We could try to come up with another explanation—another "law," that is—to explain the existence of a certain regularity. But that opens the gate for an unending appeal to more and more explanations—which amounts to an infinite regress that can never finish the job. So ultimately, we cannot evade the idea of a Lawgiver, a First Cause who gave us the laws of nature that secondary causes are based on.

[67] John Stuart Mill, *Auguste Comte and Positivism*, part I.

So we arrive back at the original idea behind the laws of nature. In spite of all that has happened in the meantime to prevent this, the laws of nature still stand tall. All attempts to discredit them fail to reduce natural laws to something that they are not. The regular rising and setting of the sun, the moon, and the stars, and the periodic motions of planets represent, in fact, grand examples of "lawful" behavior. As the late French physicist and mathematician Henri Poincaré put it, "There must be something mysterious about the normal law since mathematicians think it is a law of nature whereas physicists are convinced that it is a mathematical theorem."[68]

Here are some of the voices that still hail the laws of nature as "laws." Einstein is one of them: "Scientific research is based on the idea that everything that takes place is determined by laws of Nature."[69] Another one, Max Planck, put it this way: "There is a real world independent of our senses; the laws of nature were not invented by man, but forced on him by the natural world. They are the expression of a natural world order."[70] Or as the "father of the hydrogen bomb," Edward Teller, exclaims: "The scientist is not responsible for the laws of nature. It is his job to find out how these laws operate."[71]

Here is another, UC Berkeley physicist and Nobel laureate Charles Townes: "For successful science of the type we know, we must have faith that the Universe is governed by reliable laws and,

[68] Quoted in Mark Kac, *Statistical Independence in Probability, Analysis and Number Theory* (Rahway, NJ: Quinn and Boden Company, 1959), 52.

[69] Dukas and Hoffmann, *Einstein*, 32.

[70] Max Planck, *The Philosophy of Physics* (New York: W. W. Norton, 1936).

[71] Edward Teller, "Back to the Laboratories," *Bulletin of the Atomic Scientists* 6, no. 3 (March 1950): 71.

further, that these laws can be discovered by human inquiry."[72] Or very recently, the Templeton Prize winner and physicist Marcelo Gleiser: "There is order in the universe, and much of science is about finding patterns of behavior—from quarks to mammals to galaxies—that we translate into general laws."[73] And the list could go on.

The physicist Paul Davies seems to have changed his mind recently—perhaps to be "politically correct"—but he used to belong to the group in favor of the solid, traditional view of the laws of nature. That's when he could say, "Without this assumption that the regularities are real, science is reduced to a meaningless charade."[74] Davies expressed that same idea later on with these words: "You've got to believe that these laws won't fail, that we won't wake up tomorrow to find heat flowing from cold to hot, or the speed of light changing by the hour."[75] Given all these witnesses, we may conclude that the laws of nature really are laws, and nothing less.

Isn't it ironic that, from the beginning of the scientific era up to the middle of the eighteenth century, the discovery and first mathematical formulation of laws were used precisely in order to demonstrate the existence of a Lawgiver? From the end of the 1700s and onward, the autonomous action of those same laws was used to maintain there was no need of any Lawgiver at all, since the laws of nature worked well regardless. It certainly was not science itself that caused this change of mind.

[72] Charles H. Townes, "Logic and Uncertainties in Science and Religion," Pontifical Academy of Sciences, *Scripta Varia* 99 (2001): 300.

[73] Gleiser, "How Much Can We Know?"

[74] Davies, *Mind of God*, 81.

[75] Paul Davies, "Taking Science on Faith," *New York Times*, November 24, 2007.

Where do laws of nature come from?

Just as some scientists have difficulty calling the laws of nature what they are—laws pointing to a Lawgiver—so too do some scientists struggle to accept that the universe contains all the laws by which it operates. They probably fear that the universe, with the laws of nature it actually has, might be interpreted in its entirety as the work of God, its Creator.

In order to avoid this outcome, they have concocted a very peculiar scenario to explain our universe: the existence of "something" that spews out numerous universes quite randomly. Stephen Hawking, for instance, tried to explain this outcome with his multiverse theory—a hypothetical set of finite or infinite possible universes, including the one we live in, which together comprise everything that exists. In this scenario, the laws of nature that we find in these universes must represent nearly all possible variations and combinations. The universe we live in happens to be one of them, so the reasoning goes. And that's why the universe we live in happens to be the way it is.

Let me say first that the role of the multiverse theory seems to be a convenient alternative to the role of a Creator God in the explanation of the universe. When dealing with phenomena that appear coincidental, improbable, random, or unpredictable to us, we can always appeal to the "law of large numbers." It's a statistical law that comes in very handy when we have to deal with a highly improbable outcome—such as a particular universe with very specific laws of nature. The universe we know comes with certain laws of nature, such as the law of gravity. All of these laws could have turned out differently. In other words, the odds against a universe like ours appear to be enormous.

So the reasoning continues like this. The more universes there are—according to the statistical law of large numbers—the better the chances that a universe like ours exists. In other words, the odds of a universe like ours become better and better as we increase the number of universes, all the way up to infinity. The multiverse theory basically makes the order we find in our universe a matter of cosmic coincidence. Our universe may have a highly improbable composition, but not if there is an enormous, or even infinite, number of universes. It's like a "winning ticket." Bingo, that's how our universe came to be! No Creator needed!

Is the multiverse theory really a viable alternative to the existence of a Creator God? Some scientists think it is. Even Paul Davies has been affected by this new development, recently saying, "The laws [of nature] should have an explanation from within the universe and not involve appealing to an external agency."[76]

However, there are several reasons why the multiverse theory is not a serious alternative. First, the problem with an "infinite" number of universes is that we have no way of knowing that there is indeed an infinite number of them, not even a multitude of them. We are confined within our own universe and cannot step outside of it, except in our imagination.

Second, what we are doing here is replacing an unobservable Creator with an unobservable multitude of universes. As Stephen Barr noted, "It seems that to abolish one unobservable God, it takes an infinite number of unobservable substitutes."[77] The notion that universes are randomly spewed out is as unobservable as the idea of a Creator God with a divine plan.

Third, even if there are indeed multiple universes, the question remains where all those universes and their laws of nature come from. The question has only been shifted from one universe to many. The late philosopher and former atheist Antony Flew could not have worded it better: "So multiverse or not, we still have to

[76] Davies, "Taking Science on Faith."

[77] Barr, *Modern Physics*, 157.

come to terms with the origin of the laws of nature. And the only viable explanation here is the divine Mind."[78] The late physicist and Nobel laureate Richard Feynman used to warn his students, "I can't explain why nature behaves in this peculiar way."[79] He is right: that explanation has to come from outside of science. Paul Davies said during his more enlightened years that we would still be left, as he put it, "with the mystery why the Universe has the nature it does, or why there is a Universe at all."[80]

So we keep facing the question as to how the universe came to be, for the universe cannot be the explanation for its own existence. What, then, explains its existence?

First, the law of gravity cannot do the trick, for before the universe existed, we would have to posit laws of physics—which are ultimately the set of laws that govern the existing universe. Since the laws of nature presuppose the very existence of the universe, they cannot be used to explain it. So it would be a form of circular reasoning to claim that laws with meaning only in the context of an existing universe can generate that universe, including its laws of nature, all by themselves, before either one exists.

Second, the laws of nature could have been other than they are—they could exist or not exist; they could be this or that. It is easy to think about possible worlds or universes in which the laws of nature are radically different from those operating right here and now in our universe. Paul Davies is right: "There are endless ways in which the universe might have been totally chaotic. It might

[78] Antony Flew, *There Is a God: How the World's Most Notorious Atheist Changed His Mind* (San Francisco: HarperOne, 2008), 121–122.

[79] Richard Feynman, *The Strange Theory of Light and Matter* (Princeton, NJ: Princeton University Press, 1985), 10.

[80] Paul Davies, *God and the New Physics* (New York: Simon and Schuster, 1983), 42.

have had no laws at all, or merely an incoherent jumble of laws that caused matter to behave in disorderly or unstable ways."[81] Only a Creator God can explain why the laws of nature are the way they are. That's why, according to Isaac Newton, particles have certain forces and not others, and planetary orbits have certain parameters and not others.[82]

Third, the question arises of why there are laws of nature at all—rather than nothing. Paul Davies comments, "Over the years I have often asked my physicist colleagues why the laws of physics are what they are.... The favorite reply is, 'There is no reason they are what they are—they just are.'"[83] That's a moot reply, for these laws could have been other than they are, so they are not self-explanatory. Marcelo Gleiser is right to ask, "Why does the universe operate under these laws and not others?"[84] If there is no inherent necessity for the universe to exist, nor for the laws of nature to be the way they are, then the universe, including everything in it, is not self-explanatory and, therefore, must find an explanation outside itself. Obviously, it cannot be explained by something finite, unnecessary, and not self-explanatory—for that would lead to infinite regress—so it can only be explained by an unconditional, infinite, and necessary Being: God.

After all the above considerations, we must come to the conclusion that the universe with its laws of nature keeps pointing to God. The laws of nature that govern an existing universe can never generate that universe and bring it into existence, nor can the universe generate its own laws of nature, for that would require other laws of nature, leading to infinite regress. In short, the laws of

[81] Paul Davies, *Mind of God*, 195.
[82] Isaac Newton, *Opticks*, query 31.
[83] Davies, "Taking Science on Faith."
[84] Gleiser, "How Much Can We Know?"

nature cannot bring themselves into existence and cannot explain themselves. Therefore, the universe with its laws of nature can only find its explanation somewhere else—in a Creator God. It has often been said that "the Universe knew we were coming."[85] Yes and no: the universe did not know, but God did.

[85] Freeman J. Dyson, *Disturbing the Universe* (New York: Basic Books, 1979), 250. See also Dyson, "Energy in the Universe," *Scientific American* 224 (September 1971): 50.

3

How Physical Constants Point to God

Much of what happens in our universe depends not only on the laws of nature but also on the so-called fundamental physical constants. Some scientists see the values of these physical constants literally as "coincidences"—they just happen to be as they are. Let's see whether that can be true.

What are physical constants?

Fundamental physical constants have an unchanging, universal value that can only be determined by physical measurement—at least at this point in time. This is one of the unsolved problems of physics. Unlike the values of mathematical constants, such as π and e,[86] the values of physical constants cannot be calculated from first principles. Perhaps someday we will be able to derive their values from a more general theory that is currently not available, but as of right now, this is only a dream. So what all physical constants

[86] π (≈ 3.14) is the ratio of a circle's circumference to its diameter. This value is always the same for any circle. And e (≈ 2.72) is the base of the natural logarithm.

have in common is that their numerical values are not understood in terms of any widely accepted theory.

As Stephen Hawking has noted, "The laws of science, as we know them at present, contain many fundamental numbers."[87] In fact, those numbers are constants, a "given" in our universe. Four of them are associated with the four basic forces in nature: gravity, electromagnetism, weak nuclear interactions, and strong nuclear interactions.

Let's start with the gravitational constant, called G in physics equations. The oldest physical constant, G is an empirical physical constant used to show the force between two objects caused by gravity. The gravitational constant appears in Newton's universal law of gravitation,[88] but it also plays a role in Einstein's general theory of relativity. The measured value of the constant is notoriously difficult to measure precisely, but it is known with some certainty to four significant digits.[89]

Then, there is the strong nuclear interaction force, which works on two ranges. On the larger scale, it binds protons and neutrons together to form the nucleus of an atom. On the smaller scale, it holds quarks together to form protons and neutrons. In short, the strong nuclear force holds most ordinary matter together.[90] It is the strongest of the four fundamental forces.[91]

87 Hawking, *Brief History of Time*, 7.

88 The attractive force between two objects (F) is equal to G times the product of their masses (m_1m_2) divided by the square of the distance between them (r^2); that is, $F = Gm_1m_2/r^2$.

89 Its value is about $6.67408 \times 10^{-11} m^3 \cdot kg^{-1} \cdot s^{-2}$.

90 Its value is 2.22457 MeV (milli-electron volts).

91 At a distance of 1 femtometer (= 10^{-15} meters) or less, its strength is around 137 times that of the electromagnetic force, some 106 times as great as that of the weak force, and about 1,038 times that of gravitation.

Then, there is the weak nuclear interaction force, which is the interaction between subatomic particles (such as quarks and electrons). This force causes radioactive decay and thus plays an essential role in nuclear fission. The weak interaction takes place only at very small, subatomic distances, less than the diameter of a proton. The force is in fact termed "weak" because its field strength is typically several orders of magnitude less than that of the strong nuclear force or the electromagnetic force.[92]

Then, there is the electromagnetic force, which is more than one hundred times weaker than the strong nuclear force. Whereas the strong nuclear force tries to hold the nucleus together, the electromagnetic force tends to blow the nucleus apart, for the protons exert an electrical repulsion on each other. If the nucleus contains a large enough number of protons, their electrical repulsion will overpower the nuclear force and thus blow the nucleus apart. This means that a stable nucleus can only contain a limited number of protons. (The element with the largest stable nucleus found in the universe—uranium—has ninety-two protons.)

In addition to the constants associated with the four basic forces in nature, there are other fundamental physical constants. An example of these is the mass of the proton.[93] Protons are very stable, whereas an isolated neutron is very unstable and quickly disintegrates into other types of particles. The key for the difference in stability of protons and neutrons is their mass. Energy (E) is related to mass (m) times the speed of light squared (c^2): $E = mc^2$. The neutron has a little more mass (and thus more energy) than a proton and an electron combined. There is a general principle in

[92] The weak interaction has a value of 0.000001, compared to the strong interaction's coupling constant of 1, whereas the electromagnetic force has a coupling constant of 0.00729927.

[93] Its value is 1.672×10^{-11} kg, which gives a proton an energy equivalent of 938.272 MeV.

nature that physical systems, when left alone, seek out their lowest energy state. Sure enough, an isolated neutron will soon, within about 15 minutes on average, spontaneously turn into an electron and a proton, a process known as beta decay.

A proton is one-seventh of a percent lighter than a neutron, so it contains a bit less energy, not enough to fall apart. Therefore, a neutron can decay into a proton plus some other particles, while releasing energy in the process. But a proton cannot decay into a neutron because it does not have enough energy to do so.

Another constant, this time required for cosmology, is the cosmological constant of Einstein's equations for general relativity.[94] It has no effect on the way planets orbit stars or galaxies spiral around, but it does affect the way the universe as a whole, the cosmos, expands or contracts. It is a constant that determines how much gravitational pull is exerted by "empty space."[95] This constant is so small that physicists once thought it to be zero—but not quite. It is a constant in Einstein's equation of general relativity that, when positive, acts as a repulsive force, causing space to expand; when negative, it acts as an attractive force, causing space to contract.

Finally, there is the electromagnetic fine-structure constant,[96] which is at the basis of all the structural activity of biology, physics, and chemistry.[97]

Why are their values important?

What are we to make of these physical constants? Whereas some scientists try to tell us that we were not "meant to be here," these

[94] It has a value of approximately 10^{-120}.

[95] This refers to the relatively empty regions of the universe beyond the atmospheres of celestial bodies.

[96] It has a value of about 1/137.036, which is close to 0.007297353.

[97] Barr, *Modern Physics*, 119–137.

physical constants seem to indicate that the universe was "designed with us in mind," so that we can live in this universe and on this planet. That's why some scientists call them "anthropic coincidences," which are features that happen to be exactly what is required for the emergence of human life in this universe. By apparent "coincidence," these specific characteristics of the laws of physics allow the universe to give rise to intelligent beings like us ("*anthropos*" is the Greek word for "man"). Had any of these features been different, human life would not have been possible.

This seems to suggest that our universe is rather "fine-tuned"—as opposed to accidental and random. Since science has shown us that the conditions for life in the universe can only occur when the values of certain universal fundamental physical constants lie within a very narrow range, we must come to the following conclusion: if any of several fundamental constants were only slightly different, then the universe would most likely not be favorable to the establishment and development of matter, astronomical structures, elemental diversity, or life as we understand it.

So the question is now this: What would happen if physical constants were different from what they are, or happen to be?[98] Let's start with the gravitational constant. It has been found that its value does slightly change. This could be based on observational errors, for this constant is hard to measure. Yet what would happen if the gravitational constant were quite different? Needless to say that if G were negative instead of positive, its attractive nature would turn into a repulsive one. But what would happen if G were stronger? The apple would fall from the tree at a higher speed. If G doubled in value, then here on earth, it would take twice the effort to jump up.

[98] Paul Davies, *The Accidental Universe* (Cambridge, UK: Cambridge University Press, 1982).

But other effects could be much more serious. The moon would move nearer, and unless its orbit sped up, it would crash into the earth. The rotational speed of the earth would probably increase by a factor of a million, making a day much shorter, only some 24/1,000,000 hours long. Earth would pass about 9 to 10 percent closer to the sun than it does today.

The next thing to consider is the impact of G on the sun itself. The sun maintains a careful balance: the pressure from the hot plasma in its interior tries to tear it apart, while gravity holds it together. Again, a sudden change in the strength of gravity would disrupt this equilibrium. Hydrogen gas from the sun's surface would fall toward the core, fusing into helium and explosively releasing vast amounts of energy.

In fact, the increase in gravity, even by some 5 percent, would do its damage deep beneath the earth's surface. The earth's core bears an immense load: the entire mass of our planet, about 6.6×10^{-21} tons of rock. Thanks to the new gravitational constant, all of this rock would suddenly become 5 percent heavier. Unable to carry the extra burden, the core would collapse inward, causing the rest of the planet to fall down on top of it.

Let's move on to the strong nuclear force—the force of nature that cements nuclei together. What would happen if the strong nuclear force were stronger or weaker than it actually is? Had the strong nuclear force been *weaker* by even as little as 10 percent, it would not have been able to fuse two hydrogen atoms together to make hydrogen-2 (deuterium), a necessary step in the production of helium. The strong nuclear force would choke off the process of making the elements at the very first step, for it would be too weak to hold the nucleus of hydrogen-2 together. The prospects for life would have been dim indeed.

If, on the other hand, the strong nuclear force had been only slightly *stronger* than it is now, the opposite disaster would have occurred. It would have been too easy for hydrogen nuclei to fuse together, so the nuclear burning in stars would have gone way too fast. The entire process would have been too short for more complex forms of life such as ours to emerge.

Another example of a physical constant is the mass of the proton. As we saw earlier, the key to the difference in stability of protons and neutrons is located in their mass. A proton is one-seventh of a percent lighter than a neutron and thus contains a bit less energy. But had the proton's mass been even one percent *larger* than the neutron's, then the proton would be unstable, which means that hydrogen-1 (with only one proton and no neutron) would also be unstable. Consequently, there would be no ordinary hydrogen in the universe, and therefore no water and no organic molecules.

Paul Davies explains what would happen if protons were about 0.1 percent heavier than neutrons:

> Under these circumstances, isolated protons would turn into neutrons rather than the other way around. Some protons would be saved by attaching to neutrons. But hydrogen, the simplest chemical element, does not contain a stabilising neutron; hydrogen atoms consist of just a proton and an electron. In this backward universe, hydrogen could not exist. Nor could there be any stable long-lived stars, which use hydrogen as nuclear fuel. Heavier elements such as carbon and oxygen, made in large stars, might never form either. Without stable protons there could be no water and probably no biology. The universe would be very different.[99]

Another example is the strength of the electromagnetic force. Recall that this force is some one hundred times weaker than the

[99] Paul Davies, "The power of 1.00137841887," *Cosmos*, Autumn 2017.

strong nuclear force. Had the electromagnetic force been much *smaller*—say, only one-fifth of its current value—then there could only be some twenty-five elements in nature. Some elements essential for human life, such as calcium and iron, would not be available. On the other hand, had the electromagnetic force been *larger*, then the electrical energy packed inside a hydrogen nucleus would have been so great as to make it unstable. The "weak interaction" would then have made all the hydrogen in the world decay radioactively, with a very short half-life, into other particles. The world would have been left devoid of hydrogen, and therefore of water, which is necessary for life.

Let's return to the cosmological constant. If it were too *large*, space would expand so rapidly that galaxies and stars could not form, and if it were too *small*, the universe would collapse before life had a chance to evolve.

The above constants provide probably the best examples of anthropic coincidences. There are many other constants in physics, but they may not affect the way our universe developed. One of them is the speed of light.[100] This value (c) is used, for instance, in the famous equation $E = mc^2$. It may, in time, be tied to Planck's constant defining the smallest possible unit of space and time. Important as the value of c is, it would not be called an anthropic coincidence.

Where do physical constants come from?

As Hawking remarks about the physical constants, "The remarkable fact is that the values of these numbers seem to have been very finely adjusted to make possible the development of life."[101]

[100] Its value is 299,792,458 meters per second.

[101] Hawking, *Brief History of Time*, 125.

This fact made even the astronomer Fred Hoyle, once an outspoken atheist, exclaim, "A common sense interpretation of the facts suggests that a superintellect has monkeyed with physics, as well as with chemistry and biology."[102] There is a dawning insight here in the minds of some scientists, but they may not have the interest or the courage to think it through. Let me help them, if I may.

We said already that the values of these physical constants cannot be derived from something else, but can be determined only by physical measurement. Yet these values are such that our universe looks like a "fix." Where could they have come from? Who or what has given them these specific values? Unsurprisingly, the law of large numbers has been called in again. In the multiverse scenario, the physical constants must represent nearly all possible combinations and variations. And one of those combinations happens to be the "perfect" combination that makes human life possible in the very universe we live in.

However, as we already discussed, the multiverse theory cannot really solve the problem of how the physical constants can be so "fine-tuned." The question keeps pressing, though, as to where their values come from, if not from some wide-range theory. Well, it could be that some of them, or perhaps even all, can in time be derived from a higher-level law. For instance, at one time the boiling point of water was taken as a physical constant, but it is now considered the result of quantum mechanical laws.

However, even if it were possible to derive all constants from one single unified theory, some hoped-for super-theory, then the next question would be, of course, where that super-theory could have come from. For we are still left with the question as to where that envisioned grand unified theory came from, given the fact

[102] Fred Hoyle, "The Universe: Past and Present Reflections," *Engineering and Science* (November 1981): 8–12.

that the universe could very well have been based on other unified theories or frameworks. We would still be left, as Paul Davies puts it, "with the mystery why the Universe has the nature it does, or why there is a Universe at all."[103]

Nevertheless, there have been many attempts to come up with explanations as to why anthropic coincidences are the way they are. That's what the anthropic principle is about. It tries to explain the facts we are facing. In other words, anthropic *coincidences* are the facts, while the anthropic *principle* is a collection of many speculative hypotheses for explaining those facts.

Let's start with the weak version of the anthropic principle, WAP. It does not deny the facts. For instance, it does not deny that planet earth has some unique features. Earth is not so close to the sun that we burn up, but not so far away that we freeze solid. If it were much closer to the sun, it would be too hot to have liquid water; if the earth were much farther away, it would be too cold. If the earth were much smaller, it would not have sufficient gravity to retain an atmosphere. If it were much bigger, it would retain a lot of hydrogen in its atmosphere, which would be the wrong kind of atmosphere for life. Those are facts hard to deny. However, when it comes to the question of whether someone has "fine-tuned" the conditions here to make life possible, then the WAP defenders say, "Not necessarily so."

Stephen Barr makes a strong point against the WAP by stressing that, if the order of the universe were the result of mere chance, as the WAP asserts, we still need to face the fact that all constants are interrelated and interdependent in such a way that changing only one of them would mean changing practically all of them, thereby changing the characteristics of the entire physical world we know. Besides, we have not explained yet how this orderliness could be

[103] Davies, *God and the New Physics*, 42.

so "perfect" that the laws of nature apply to everything anywhere in the universe, without any exceptions. His point is that we find *perfect* order and lawfulness in our universe:

> Among all the logically possible Universes, ones that have the perfection of order and lawfulness that ours displays are highly exceptional, just as among all possible rocks, a perfect gem that has absolutely no flaws in it is almost infinitely unlikely. Why doesn't our Universe exhibit occasional departures from its regularities—the regularities we call the laws of physics—just as gemstones have occasional departures from their regularities? No answer to this is possible.[104]

The WAP basically glosses over these questions. So why is it still popular among scientists? Well, it seems to be an explanation that fits within a scientific framework, for it works with concepts such as randomness that have a scientific allure. But that doesn't change the fact that the WAP cannot be scientifically proven itself—it remains a mere hypothesis. Perhaps the best assessment of the WAP is that it gives scientists an easy alibi to avoid looking into its much stronger version, the SAP.

Unlike the WAP, the SAP attempts to explain why such improbable events did occur, rather than just state that improbable events must have occurred based on the law of large numbers. It acknowledges that there is indeed something "remarkable" about the values behind the physical constants. It also acknowledges the "amazing" fact that there is such a close match between what physics determines and biology requires. Could that be mere coincidence, too? We cannot ignore the fact that all of these far-fetched coincidences happened together and so perfectly in unison—which

[104] Stephen M. Barr, "Anthropic Coincidences," *First Things*, June 2001.

is hard to call a coincidence. The theoretical physicist Freeman Dyson put it this way: "As we look out into the Universe and identify the many accidents of physics and astronomy that have worked together to our benefit, it almost seems as if the Universe must in some sense have known that we were coming."[105] There is too much coincidence in the coincidences combined. Can the existence of so many coincidences be mere coincidence?

The SAP at least accepts that there is fine-tuning, but this does not mean, of course, that the fine-tuning is done by God, who has an end or goal in mind for the universe and for humanity, and thus has consciousness and foresight. Yet it seems to me that all the previous "solutions" are in essence nearly desperate moves to avoid the only sensible alternative—which is that the universe was designed by a Creator God who had us in mind. According to that scenario, we were predestined to emerge, and planet earth was predestined to become our home. It is hard *not* to accept that, behind it all, there must be a divine plan, the work of a Divine Mind. As the physicist John A. Wheeler put it, "A life-giving factor lies at the center of the whole machinery and design of the world."[106] Earlier, we quoted the physicist Paul Davies who said at one point, "There must be an unchanging rational ground in which the logical, orderly nature of the Universe is rooted."[107]

All of this makes the quest for a Designer not only highly relevant but even inescapable. I see no other solution than that the order of the universe is the result of a design that was implemented—actually conceived, invented, and decreed—by a Divine

[105] Dyson, *Disturbing the Universe*, 250. See also Dyson, "Energy in the Universe," 50.

[106] John A. Wheeler, "Foreword," in *The Anthropic Cosmological Principle*, by John D. Barrow and Frank J. Tipler (Oxford, UK: Clarendon Press, 1986), vii.

[107] Davies, "What Happened before the Big Bang?," 10–12.

Designer who had a divine plan in His Divine Mind. Not only is God the Lawgiver of the laws of nature, but He is also the Engineer of the physical constants. In other words, our universe is a work of "cosmic engineering." So it shouldn't surprise us that God has often been described as an Architect, an Artisan, or a Workman (*Deus Faber*). As the book of Wisdom says about God, "Thou hast arranged all things by measure and number and weight."[108]

[108] Wisd. 11:20.

4

How a Grand Unified Theory Points to God

"Grand theories" have always been appealing. Many of us searched for them, are still searching for them, and will keep searching for them. They are like dreams we would like to come true. They are seen as giving us a better understanding of our world, and perhaps also more power. Is the "Grand Unified Theory" (GUT) one of them?

Why a Grand Unified Theory?

Physics is far from complete. Whether it ever will be is another question, but it is clearly incomplete at the present time. One of the reasons why it is incomplete is that in order to unify the three non-gravitational forces—the electromagnetic force, the strong nuclear force, and the weak force—we would need an "umbrella" theory that unifies *all* of nature's forces as manifestations of *one* single, all-encompassing force. Scientists have made progress in unifying the three quantum forces, but gravity has remained a problem in every attempt. Physicists have tried to "quantize gravity" by positing a particle, the graviton, that carries gravity in the same way that photons (light) carry electromagnetism. But so far to no avail. Without a unifying umbrella theory, physics would certainly be incomplete.

So we don't have a theory of gravity that is consistent with both the theory of relativity and quantum theory. Here is the problem. We do have theories about space, time, and gravitation (Einstein's general relativity theory) as well as theories about particles, electromagnetism, and quarks (quantum theory). However, these two sets of theories can be at odds with each other.

Quantum physics, on the one hand, assumes a framework of space and time for any processes to take place; it deals also, but not exclusively, with small particles when gravitation can practically be neglected. The relativity theory, on the other hand, states that curvature of space and time is caused by matter; it holds for situations where gravitation is considerably more effective.

In other words, quantum effects are typically (but not always) more important for small systems than for large ones. Now the following problem arises. When we go from large systems to small systems, distances decrease while density increases. At some point, distances become so short and density so concentrated that quantum theories and gravitation theories need to be combined. However, this requires a theory that unifies gravity with other non-gravitational forces. And that's what is missing.

The defining postulates of both Einstein's theory of relativity and quantum theory are indisputably supported by rigorous and repeated empirical evidence. However, while they do not directly contradict each other theoretically, they are resistant to being incorporated within one cohesive model. There is no known quantum theory of gravity for things smaller than, say, common dust particles, so gravity isn't even part of the equation. There is an enormous gap left. Now the goal of a GUT is to bridge this gap. This could be done if we knew the smallest building blocks in physics.

Finding the smallest building blocks, though, is only the first step in devising a GUT. The next step is understanding the forces that govern how the building blocks interact. For three fundamental forces in nature, physicists know already that each of the known subatomic forces has an associated particle or particles that carry that force: the gluon carries the strong force, the photon governs the electromagnetic force, and the W and Z bosons control the weak force. Taken together, these building blocks and forces make up the so-called "Standard Model" of physics. Using quarks and leptons and the known force-carrying particles, one can build atoms, molecules, planets, and, indeed, all of the known matter of the universe.

Currently, the Standard Model has twelve matter particles (six quarks and six leptons) and four forces (electromagnetism, gravity, and the strong and weak nuclear forces). Finding a smaller building block will be difficult, because that requires a more powerful particle accelerator than scientists have ever been able to build. Yet there is reason for hope, because unification is nothing new in physics. Think of the following cases in history. Newton's laws united earthly and heavenly bodies. Electromagnetism united electrical, magnetic, and optical phenomena. Quantum theory united particles and waves. Einstein's four-dimensional concept of space-time united space and time. And now a still unknown GUT might someday unite, unify, and combine Einstein's gravity and Planck's quantum theory. A mathematical unification of the forces is reasonable, and it may succeed because the universe seems to be susceptible to being unified.

What would happen if this goal could be achieved? The search is on, and may still go on for quite a while. All we can say at this point is this: there have been trials on the "drawing board" to combine Einstein's gravity and Planck's quantum theory. The best of these trials is *superstring* theory. We do not know whether superstring theory is the true theory, and we may never know, but many physicists have good reasons to assume it is.

We won't go into details, for it's a very complicated theory with very complex mathematical equations, entirely beyond my expertise. Let me just say this. Superstring theory postulates that the smallest building block of the universe isn't a particle, but rather a small and vibrating "string." In the same way a cello string can play more than one note, the different patterns of vibrations make for the different quarks and leptons.

If there is indeed such a thing as a GUT, then everything would be so tied to everything else that nothing could be changed without destroying the whole structure of the theory. This means that in the ultimate theory, as Stephen Barr puts it, "it may turn out that everything has to be just as it is."[109] Everything would be uniquely determined. Interestingly enough, Einstein once said, "What I am really interested in is whether God could have made the world in a different way."[110] If there is a GUT, that theory would uniquely determine the universe as it is. That could be seen as the "thought" God had in mind for this universe.

Can science ever accomplish this goal of finding a GUT? Only time can tell; it's an empirical question. So we need to wait. But there is more to consider about this theory.

What a Grand Unified Theory is not

Some people call a GUT, regardless of whether it exists or not, a "Theory of Everything" (TOE). But that's a serious misunderstanding. A GUT is not a TOE. Why not?

A GUT, if it exists, would at best offer us only a physical explanation about physical phenomena, not a complete theory of *everything*. It could not, for instance, explain why some people believe it and some do not. There simply is no science of "all there is." One cannot give science, nor any of its theories, the metaphysical power it does not possess.

[109] Barr, *Modern Physics*, 142.
[110] Quoted in ibid.

But what makes some believe that all phenomena are physical phenomena? Physics, for its part, discloses only one of the many aspects of the material world, its physical aspect, thus providing physical answers to physical questions phrased in terms of physical causes and effects. Physics may be everywhere, but it's surely not all there is. Biology offers another perspective, providing biological answers to biological questions in terms of physical causes and biological functions. And religion offers yet another perspective. But there is no science of "all there is."

The GUT does not explain "everything." Where in this theory is life, where are cells and organs, plants and animals, brains and consciousness, love and hate, God and creation, and so on and so on? Would the GUT ever be able to connect phenomena as disparate as quantum mechanical states and the songs of birds and the thoughts of humans and the culture of a society and the proofs of God's existence? Apparently, the TOE has much broader pretentions than the GUT. A TOE may be one of those pursuits humanity has been dreaming of for centuries. But it is hard to see how a TOE could ever be possible, let alone be found.

A TOE pretends to answer all the questions we have in life, leaving nothing unanswered. But a GUT can answer only our physical questions. Well, not all questions are physical questions. So someday there may be a "Grand Unified Theory of Physics" but that is not the same as a "Grand Unified Theory of Everything." Therefore, we cannot equate a GUT with a TOE. A TOE is arguably one theory too far.

Can a Grand Unified Theory replace God?

If we ever find a GUT, doesn't that mean that God has nothing to do anymore, because then the GUT is in charge of all that happens in this world? No, this doesn't follow for several reasons.

First of all, even if all physical constants and laws of nature can someday be derived from one GUT, we can't say that all physical constants and laws of nature are determined by this unified theory—instead of by God. For we are still left with the question as to where that GUT came from, given the fact that the universe could very well have been based on other theories or frameworks. No GUT is self-explanatory.

We would still be left, as Paul Davies puts it, with the question of why there is a universe at all. Stephen Hawking seems to be vaguely aware that there is indeed a problem here, when he says, "Even if there is only one possible unified theory, it is just a set of rules and equations. What is it that breathes fire into the equations and makes a Universe for them to describe?"[111] The answer to that question leads us almost inevitably to a divine plan in the Divine Mind of a Creator—which would mean we were predestined to be here. If there is indeed a GUT, it would still point to God. A world without God is incomplete, missing a vital "element" beyond the reach of a GUT.

Second, a GUT might make everything uniquely determined, but only in the world of physics. But fortunately, I would say, there is more to the world than physics. How could anyone ever prove, for instance, that physics determines what human beings think and do? There is no proof for that, but only a metaphysical conviction called reductionism. It is an ideology that has become the source of slogans such as "Life is nothing but chemistry and physics" (James Watson); "Humans are nothing but a speck of dust" (Carl Sagan); "Humans are nothing but a pack of neurons" (Francis Crick); "The brain is nothing but a 'meat machine'" (Marvin Minsky); "Humans are nothing but glorified animals" (Charles Darwin); and "Humans are nothing but a bundle of instincts" (Sigmund Freud).

[111] Hawking, *Brief History of Time*, 174.

In all these cases, the words "is" and "are" feature as key words. However, one could argue against claims like these that, within the setting of a biological model, human beings may indeed be seen as "*only* DNA," but in reality they are "*also* DNA." There is more to human beings than the physics and chemistry of DNA. Seen this way, the slogan "Humans are nothing but DNA" makes no sense. It has reductionism written all over it.

Reductionism reduces everything to its simplest elements. It is rooted in the metaphysical doctrine of materialism—"Matter is all that matters." Although many scientists are materialists, materialism is not a scientific conclusion, but merely a philosophical opinion, an ideology. Defending that a GUT is also a TOE is basically a form of materialism, because it amounts to asserting that "everything" is physical, and only physical, so a GUT would necessarily be a TOE, too. True, scientists have been very successful in explaining things by analyzing them into smaller and smaller parts and particles—which is the "piecemeal" approach scientists are known to excel at. It is certainly possible to study human beings by studying their molecules, atoms, and subatomic particles, but that does not mean human beings are ultimately "nothing but" those parts.

As the late Austrian physicist and Nobel laureate Erwin Schrödinger stated in his famous book *What is Life?*,[112] life and living phenomena cannot be reduced to the principles that physics has identified without losing the essence of life itself. This idea has been rephrased many times—for instance: "When you try and take a cat apart to see how it works, the first thing you have on your hands is a non-working cat."[113] Or this one: "When

[112] Erwin Schrödinger, *What Is Life?* (Cambridge, UK: Cambridge University Press, 1944).

[113] Douglas Adams, *The Salmon of Doubt: Hitchhiking the Galaxy One Last Time* (New York: Random House, 2002), 135–136.

you cut apart a living frog to 'see what makes it tick,' the 'ticking' disappears, but all the pieces of the frog are still there."[114] What remarks like these suggest is that there might be more to life than what science tells us. That's how the door to God keeps opening up.

Third, the idea that a GUT determines that "everything has to be just as it is" doesn't mean that God had no choice but to implement a GUT. Of course God had a choice. God could have chosen a different unified theory, or a non-unified theory, even no theory at all. There would actually be an infinite number of different mathematical structures that could serve as the laws and constants of some hypothetical universe. There could, for instance, be an infinite number of dimensions and an infinite number of possible particles. To put it briefly, the universe does not have to be the universe it is. If there is indeed a GUT, it would still point to God.

What remains standing is that the physical order we observe in this universe appears to be amazingly consistent. It follows laws of nature that can be unified into higher-level theories with great elegance and harmony—and perhaps someday even into one single unified theory. This is true as long as we are aware of the difference between a GUT and a TOE.

Perhaps what we discussed here is well summarized by the physicist Marcelo Gleiser during an interview:

> It is impossible for science to obtain a true theory of everything. And the reason for that is epistemological. Basically, the way we acquire information about the world is through measurement. It's through instruments, right? And because

[114] Monica Anderson, "Reduction Considered Harmful," *H+ Magazine*, March 31, 2011.

of that, our measurements and instruments are always going to tell us a lot of stuff, but they are going to leave stuff out. And we cannot possibly ever think that we could have a theory of everything, because we cannot ever think that we know everything that there is to know about the universe.[115]

[115] Gleiser, "Atheism Is Inconsistent."

5

How The Big Bang Points to God

The English astronomer and mathematician Fred Hoyle is credited with coining the term "Big Bang" during a 1949 radio broadcast. But the concept behind it had already been suggested in 1927 by the Belgian priest, astronomer, and physicist Georges Lemaître of the Catholic University of Louvain.

Currently, there is little doubt among cosmologists that we live in the aftermath of a giant explosion that occurred some fourteen billion years ago. The Big Bang theory is now the prevailing cosmological model that explains the early development of the universe. Although nothing in science is final, cosmologists tell us that the Big Bang theory is still the latest and best we have at the moment; it remains standing for now until further notice.

The universe started with the Big Bang

Our universe most likely started with the Big Bang, some fourteen billion years ago. Edwin Hubble discovered in 1929 that the distances to faraway galaxies were generally proportional to their redshifts.[116]

[116] Edwin Hubble, "A relation between distance and radial velocity among extra-galactic nebulae," *PNAS* 15, no. 3 (March 15, 1929): 168–173.

"Redshift" is a term used to describe situations when an astronomical object is observed to be moving away from the observer, such that emission or absorption features in the object's spectrum have shifted toward longer wavelengths—red, that is. Hubble's observation was taken to indicate that all very distant galaxies and clusters have an apparent velocity directly away from our vantage point—the farther away, the higher their apparent velocity.

Two years earlier, Georges Lemaître made this very suggestion. In 1931, Lemaître went even further and suggested that the evident expansion of the universe, if projected back in time, meant that the further in the past, the smaller the universe was, until at some finite time in the past, all the mass of the universe was concentrated in a single point—a "primeval atom," in Fr. Lemaître's own words. That's where and when the fabric of time and space must have started. Time and space are like the "fabric" or "matrix" of the universe.[117]

According to the Big Bang theory, the universe was once in an extremely hot and dense state, which expanded rapidly. This rapid expansion caused the universe to cool and resulted in its present continuously expanding state—not with galaxies moving through space, but rather the space between the galaxies being stretched out. Once it had cooled sufficiently, its energy was allowed to be converted into various subatomic particles, including protons, neutrons, and electrons. Giant clouds of these primordial elements would then coalesce through gravity to form stars and galaxies, and the heavier elements would be synthesized either within stars or during supernovae.

Interestingly enough, the ninety-two elements we find on planet earth can be found all over the universe, indicating a common

[117] Georges Lemaître, "The Beginning of the World from the Point of View of Quantum Theory," *Nature* 127, no. 3210 (May 1931): 706.

origin. Stars are prodigious factories of elements. So the presence of certain elements gives us an indication about the age of stars and planets. Since iron, for instance, is produced very gradually inside the "furnace" of a star, lack of iron shows that a system (including its stars), didn't have enough time to produce iron, and thus must date from a time before elements such as iron became abundant.

From Big Bang to planet earth

A combination of observation and theory suggests that the first galaxies were formed about a billion years after the Big Bang. Since then, larger structures have been forming, such as the Milky Way galaxy. Nebulous blobs of gas formed the starting point of our galaxy.[118] They consisted merely of hydrogen and helium (and a bit of lithium), which were created in the Big Bang.

Our galaxy, the Milky Way galaxy, has been described as an "exceptionally quiet" spiral galaxy.[119] Its magnetic field is relatively weak, yet strong enough to prevent the collapse of its spiral structure. Its disk is dense enough to sustain its surprisingly beautiful shape—which is so common among galaxies that the universe almost seems to delight in building them.

Part of the Milky Way galaxy is our solar system—composed of the sun (a star) and all the planets around it, including earth. Several billion years ago, the solar system was nothing but a cloud of cold dust particles swirling through largely empty space. This cloud of gas and dust started to collapse as gravity pulled everything

[118] Cristina Chiappini, "The Formation and Evolution of the Milky Way," *American Scientist* 89, no. 6 (November/December 2001): 506–515.

[119] François Hammer et al., "The Milky Way, an Exceptionally Quiet Galaxy: Implications for the Formation of Spiral Galaxies," *Astrophysical Journal* 662 (June 2007): 322–334.

together, forming a huge spinning disk. As it spun, the disk separated into rings. The center of the disk grew to become the sun, and the particles in the outer rings turned into large fiery balls of gas and molten liquid that cooled and condensed to take on solid form. About 4.5 billion years ago—at 4:17 p.m. projected on the scale of a day, and on September 3 on the scale of a year—they began to turn into the planets that we know today. Planet earth was one of them.

For earth to be the way it is and to be able to support life, including intelligent life, numerous factors had to be a certain way. Earth must have an orbit around the sun that is closer to a circle than an elongated ellipse. It must be a volcanic planet that generates the right kind of gases. It also must generate a magnetic field to deflect the most harmful rays coming from the sun. It must have a good balance of land and water. Its continents must be rather evenly distributed to limit deserts and other weather extremes. Earth must also have a steady rotation on a stable axis that permits seasons and a proper distribution of rain. It must have also a certain relationship with the moon so that tides are steady. And the list goes on and on.

How is this highly improbable combination of numerous factors possible? Of course, the law of large numbers has been used to explain the statistical improbability of this rare combination. Since there are so many planets in so many galaxies, scientists think they have reason to assume that there can be at least one planet where the proper, improbable conditions for life have been realized. The idea is basically simple: the more planets there are, the better the chances are that there is a planet with the conditions we find on planet earth. Given an almost infinite number of planets in the universe, some of them are hot, some cold, some big, some small, and so on. They undoubtedly span a wide range of physical and chemical conditions. It seems inevitable that some of them would happen to have the right conditions for life. Right?

Perhaps not. There is just a kind of God-given perfection to earth that is hard to replicate statistically—to have so many coincidences together can hardly be a coincidence. The outcome is so unlikely and improbable that this process seems to have been "orchestrated" in some way from somewhere else. This process was partly, perhaps even mostly, based on laws of nature and the values of physical constants, which were implemented by the Creator of the universe, as we discussed earlier.

Let's explain this with one law of nature in particular: the law of gravitation.[120] Everything in the universe, from the tiniest atom to the largest planet, moves in a precise trajectory governed by the exact mathematical equations of the laws of nature. Had the law of gravitation been different, the universe and planet earth could not be the way they are right now, and we would most likely not be here.

This idea is certainly not new. William Paley, already in 1802, pointed out that if the law of gravity had not been an inverse square law (= $1/\text{distance}^2$), then the earth and the other planets would not be able to remain in stable orbits around the sun.[121] Instead, there is a balance between the force of gravity and the centrifugal force—otherwise planets would fly off to infinity or plunge into the stars they were orbiting. Without a law based on the inverse square of distance, the orbits of the planets in the solar system—and anywhere else for that matter—would be much more complicated than ellipses, and in most cases would not form closed curves at all.

Paley also remarked that if space had not been three-dimensional but, for example, four-dimensional, gravity would have decreased relative to the inverse *cube* ($1/d^3$) of the distance, rather

[120] Remember that the attractive force between two objects (F) is equal to G times the product of their masses ($m_1 m_2$) divided by the square of the distance between them (r^2); that is, $F = Gm_1 m_2/r^2$.

[121] William Paley, *Natural Theology*, 12th ed. (London: J. Faulder, 1809), chap. 22, 390–391.

than the inverse *square* ($1/d^2$). That change in the character of gravity would also have made planetary orbits unstable, and the continued existence of the solar system would have been in constant jeopardy.

We take it for granted that space is three-dimensional, but it does not have to be that way—it's neither a logical nor a metaphysical necessity, but rather an empirical fact. Had this fact been different, then it would have been impossible for planets to orbit stably around stars: they would either have plummeted into stars or flown off into space. On the other hand, had there been fewer than three dimensions, as Stephen Barr notes, complex organisms would have been impossible for quite a different reason. Complex neural circuitry, as is needed in a brain, would not be possible in two dimensions. If one tries to draw a complicated circuit diagram on a two-dimensional surface, one finds that the wires have to intersect each other many times, which would lead to short-circuits.[122]

Why are these things the way they are? Are they just mere coincidences? Can it be a coincidence that there are so many coincidences together? Or may we say instead that God has prevented the "wrong" possibilities from occurring so we could arise in His creation? I don't see why not. God created what's best for us and for all his other creatures. In other words, there is so much in our universe that points to God. St. Paul told the Christians in Rome that we can see the invisible God through what we see around us: "Since the creation of the world his invisible nature, namely, his eternal power and deity, has been clearly perceived in the things that have been made."[123] All of them point to God, both individually and combined.

[122] Barr, *Modern Physics*, 133.

[123] Rom. 1:20.

The Big Bang could not start itself

Where did the Big Bang come from? It is deceivingly attractive to think that the Big Bang started itself. Stephen Hawking is one of those who believe this. He talks about the Big Bang in terms of what he calls a "spontaneous" creation: "Because there is a law such as gravity, the Universe can and will create itself from nothing. Spontaneous creation is the reason there is something rather than nothing."[124] This, in turn, made the late astrophysicist Carl Sagan exclaim in the preface to one of Hawking's books that such a cosmological model has "left nothing for a creator to do."[125]

Hawking may be a good scientist, but he is a poor philosopher. Einstein used to say, "It has often been said, and certainly not without justification, that the man of science is a poor philosopher."[126] Hawking is an example. His idea of "spontaneous creation" is sheer philosophical magic. For something to create itself spontaneously, it would have to exist before it came into existence—which is logically and philosophically impossible. How could the universe "create itself" from nothing—not to mention cause to be itself? The law of gravity cannot do the trick, for even if the universe could create itself, if even possible, we would first have to posit laws of physics—which are ultimately the set of laws that govern the existing, created universe. The law of gravity is certainly not nothing!

So we have a *logical contradiction* here, saying that laws that have meaning only in the context of an existing universe can generate that universe, including the laws of nature, all by themselves

[124] "Stephen Hawking and Leonard Mlodinow, *The Grand Design*," *Times Eureka* 12 (September 2010): 25.

[125] Introduction to Hawking, *Brief History of Time*.

[126] Albert Einstein, "Physics and Reality," *Journal of the Franklin Institute* 221 (March 1936): 348–382.

before either exists. Besides, Hawking's explanation of "spontaneous creation" is a perfect example of *circular reasoning*: you can't have a universe without its being created, you can't have spontaneous creation without physical laws, and you can't have physical laws without a universe. That's how we keep circling around without getting anywhere. To have anything—a universe, a law of gravity, or anything—we need to have something else first: the Creation by the Creator.

To avoid this muddle of misconceptions, we should learn from the vital distinction St. Thomas Aquinas makes between producing (*facere*) and creating (*creare*).[127] Most people use these two terms interchangeably, but Aquinas advises us to separate them for clarity's sake. Here is why.

Science is about "producing" something from something else—it is about changes taking place in this universe. Creation, on the other hand, is about "creating" something from nothing—which is not a change at all, certainly not a change from "nothing" to "something." Aquinas puts it this way: "To create is, properly speaking, to cause or produce the being of things."[128] In other words, God the Creator doesn't just take preexisting stuff and fashion it, as does the Demiurge in Plato's *Timaeus*. Nor does he use something called "nothing" and then create the universe out of that. Rather, God calls the universe into existence without using preexisting space, matter, time, or anything else.

Neither is Creation something that happened long ago nor is the Creator someone who did something in the distant past, for the Creator does something at all times—by keeping our contingent world in existence. Our world is contingent, which means it does not have to exist, but came into being and will pass away in time.

[127] Aquinas, *De Symbolo Apostolorum* 4, 33.
[128] Aquinas, *Summa Theologica* I, q. 45, art. 6.

Creation is not a change; it's a cause, but of a very different, indeed unique, kind—a First Cause. The book of Wisdom puts it this way: "How could a thing remain, unless you willed it; or be preserved, had it not been called forth by you?"[129] To paraphrase Shakespeare's *Hamlet*, Creation is about "to be or not to be."

Whereas the universe may have a beginning and a timeline, Creation itself does not have a beginning or a timeline. Instead, Creation actually makes the beginning of the universe and its timeline possible. William E. Carroll is right to stress that we should never confuse temporal *beginnings* with theological *origins*. In his own words, "We do not get closer to creation by getting closer to the Big Bang."[130] In other words, we do not get closer to our origin by getting closer to the beginning of this universe. Once we lose sight of this important distinction, we are in for a serious mix-up with dangerous consequences.

It was in 1215 that the Fourth Lateran Council taught that, at the beginning of time, the universe was "created out of nothing."[131] The Council was actually saying that *creating* something "out of nothing" does not mean *producing* something out of nothing—which would be a conceptual mistake, for it treats "nothing" as some kind of thing. In contrast, the Christian doctrine of Creation "out of nothing" (*ex nihilo*) claims that God made the universe without making it out of anything. Creation has everything to do with the philosophical and theological question as to why things exist at all, before they can even undergo change.

What should we conclude from this? Creation—but certainly not the Big Bang—is the reason why there is something rather

[129] Wisd. 11:25, NABRE.

[130] William Carroll, "The Genesis Machine: Physics and Creation," *Modern Age* (Winter/Spring 2011).

[131] Denzinger, *Enchiridion*, 428 (355).

than nothing (including something such as the law of gravity). Science is about "producing" something from something else; religion is about "creating" something from nothing. So the Big Bang theory is about the *beginning* of the universe—about how physical interactions came about—whereas faith in Creation is about the *origin* of the universe—about where the universe has come from and how it completely depends on God for its existence.

The Big Bang theory did not replace God nor make Him redundant. Georges Lemaître, who launched the Big Bang theory, once spoke about the God of the Big Bang as the "One Who gave us the mind to understand him and to recognize a glimpse of his glory in our Universe which he has so wonderfully adjusted to the mental power with which he has endowed us."[132] Just as the Grand Unified Theory keeps pointing to God, so does the Big Bang theory. Without God, both of them are missing the point.

[132] Georges Lemaître, *The Primeval Atom* (New York: D. Van Nostrand, 1950), 55.

6

How Genetics Points to God

The field of genetics studies how genes affect the appearance of living beings, and how their DNA is transmitted to the next generation. Thanks to DNA, we are connected with a long, long line of ancestors, so each one of us didn't have to be invented anew from scratch.

The basics of genetics

In the 1860s, the Augustinian abbot Gregor Mendel made the discovery that genes—the basic units of genetics—could be transmitted from one generation to the next. Then, in 1910, Thomas Hunt Morgan proved that genes do so by being carried on chromosomes. The human genome—the genetic material of the species *Homo sapiens*—consists of 23 pairs of chromosomes. Cells usually carry one copy of the chromosomes 1–22 from each parent, plus an X chromosome from the mother, and either an X or Y chromosome from the father, which makes for a total of 46 chromosomes, or 23 pairs. These chromosomes carry genes.

Then, in the 1940s, it was discovered that genes contain a certain molecule called DNA (deoxyribonucleic acid). In other words, the cell carries genes in the form of chromosomes, and genes are made up of DNA. It was in 1953 that James Watson and Francis Crick published the structure of DNA with their "double helix" model. From then on, developments followed a speedy course. Soon, it was found out that the DNA of a gene provides the blueprints for a protein.

The building blocks of proteins are amino acids, which come in 20 varieties. The building blocks of DNA are nucleotides (the bases A, C, G, and T in the DNA), which come in units of three. Each unit contains the code for one of those 20 amino acids. So a gene has the building instructions for a specific protein. The human genome, in total, contains the building instructions of all the proteins needed for a human body.

Proteins differ from one another primarily in their sequence of amino acids, dictated by the DNA of their genes. Some of these proteins have a structural function (such as actin and myosin in muscles, or keratin, collagen, and elastin found in bones and skin). Others have a catalytic function and are also called enzymes (examples are amylase in saliva and DNA-polymerase in every cell). Acting as catalysts, enzymes speed up reactions without being used up in the process, so only small quantities of enzymes are required to carry out a reaction.

The process of assembling proteins based on the DNA code can be simplified as follows. The step from DNA to protein involves another molecule called RNA, very similar to DNA. First, the DNA of a gene makes a matching RNA version of its code—a process called transcription—and then the RNA string determines which amino acids are being strung together into a protein—a process called translation. This way, DNA is indirectly in control of the proteins a cell produces and how they work.

Genes and DNA carry information

Geneticists use a strikingly distinctive vocabulary foreign to physics and chemistry. They quite nonchalantly say that DNA is a genetic "code" that contains the genetic "information" for a living being. Articles written by geneticists are laden with terms such as "code," "message," "program," "information," "blueprint," and so many more. Of course, we can argue that geneticists don't mean what they say, but if they do, we have to take them seriously. Let's give them the benefit of the doubt. The easiest way out is to say that they are using metaphorical language—as a figure of speech. But there must be more to it.

Basically, geneticists are saying that genetics cannot be studied from a purely physicochemical perspective. DNA is not simply a chemical molecule but a biological structure that must be understood and explained in terms of a code or program. In other words, to explain the working of DNA, we cannot look to chemistry and physics alone, but to information technology as well.

Crucial in information theory is the separation of content from the vehicle that transports it. If it is true that the same information can be transported by different vehicles—such as pen strokes on paper, currents in computers, or DNA in the cell—then the conclusion must be that information is not identical to its carrier. The message of a code can be analyzed in terms of its particular physical and chemical elements, but apparently there must be more to it than those elements.

From a physicochemical viewpoint, any order of nucleotides in the DNA sequence is possible, but this is not true from a biological point of view. Therefore, physics and chemistry cannot specify which order will in fact convey information. To explain the difference, we could use the analogy of writing information on paper. If the ink or the letters dictated the content of the message, we would be severely limited in terms of the information we could communicate. The late physical chemist Michael Polanyi expressed this as follows: "[As] the arrangement of the printed page is extraneous to the chemistry of the printed page, so is the base sequence in a DNA molecule extraneous to the chemical forces that work in a DNA molecule."[133]

This explains why DNA would not be able to store or convey any information at all if its sequence were fully predetermined by physical and chemical bonds. DNA may be a carrier of information, but in itself, it is not information. Any arrangement of nucleotides in DNA is compatible with the laws of physics and chemistry, and must be so; otherwise, it could not exist at all. Put differently, the genetic "words" of the DNA code are no more dictated by the chemistry of their nucleotides than the words in a newspaper are determined by the chemistry of their carrier. As George F. Gilder of the Kennedy Institute at Harvard University puts it, "Information is defined by its independence from physical determination: If it is determined, it is predictable and thus by definition not information."[134]

If the genetic terminology in terms of information is indeed not merely metaphorical language, then we may be able to understand better several things. First, there are "synonyms" in the "words" of DNA. Because the coding unit of DNA, a codon, is three nucleotides long, which makes for sixty-four possible combinations (4^3), there is room for "synonyms" to code for twenty amino acids (for example, the codons GCA, GCC, GCG, and GCU all specify the same amino acid, alanine).

Second, the "meaning" of "words" in DNA may depend on their context. The Harvard geneticist Richard Lewontin likes to use the DNA sequence of the nucleotides GTAAGT as an example. Usually the code is "read" as a two-codon instruction to insert the amino acids valine and serine into a growing protein chain. However, sometimes this very same sequence is "read" as

[133] Michael Polanyi, *Knowing and Being: Essays by Michael Polanyi* (Chicago: University of Chicago Press, 1969), 229.

[134] George Gilder, "Evolution and Me," *National Review*, July 17, 2006, 29s.

a code that regulates the expression of a neighboring gene, and at other times, it acts as a "blank" separating two different DNA sequences.[135] Apparently, DNA is a code that works in a context.

Third, there can be "misspelled words" in DNA, which may affect their "meaning." Code fragments that were damaged by mutations can actually become "nonsense" code—one of those other words foreign to physics and chemistry. Nothing in physics and chemistry qualifies as nonsense, but in genetics, it can be. "Misspelled" parts of DNA may lose their information and become gibberish.

Is DNA the secret of life?

Given the important and unique role of DNA in carrying the genetic information that living beings need for their existence, it should not come as a surprise that many have come to believe that DNA is "the secret of life." That statement is certainly a popular mantra, but it needs to be treated with great care. Why?

First of all, DNA in itself—for instance, in a test tube—cannot do anything. It's like a dead piece of text that needs to be read to make sense. DNA needs the extended machinery of the cell to make its information available. This means that DNA isolated in a test tube is unable to use its information. Think of viruses, which are essentially pure DNA or RNA; their DNA or RNA cannot do anything until they penetrate, like a Trojan horse, the interior of a "living" cell where they utilize its machinery. So "the secret of life" is certainly not confined to the role of DNA.

Second, DNA cannot even copy itself on its own. It is not capable, as many still believe, of self-replication—that is, making

[135] Richard Lewontin, *Biology as Ideology* (New York: Harper Perennial, 1993), 68. Essay originally printed in *New York Review of Books*, May 28, 1992.

DNA-copies for new cells all by itself. If DNA is put in the presence of all the pieces that can be assembled into new DNA, but without any protein machinery, nothing happens. It is actually the presence of many other components that makes sure old DNA strands are replicated into new strands. This process is analogous to the production of copies of a document by an office copy machine—a process that would never be described as "self-replication."

Third, even though DNA does specify the order of amino acids in proteins, the end product may not be a functional protein yet. To become a protein with physiological and structural functions, it must be folded into a three-dimensional configuration that is based only partly on its amino acid sequence, but is also determined by the cellular environment and by special processing proteins. DNA has effectively a one-dimensional configuration, while proteins are three-dimensional structures. There is no direct line from one to the other. So the DNA information is far from complete.

The industrial production of insulin makes a case in point. Recently, the DNA coding sequence for human insulin was inserted into bacteria, which were then grown in large fermenters until a protein with the amino-acid sequence of human insulin could be extracted to fight diabetes. However, the first proteins harvested through this process did have the correct sequence, but were physiologically inactive. Apparently, the bacterial cells had folded the protein incorrectly. Somehow, DNA itself does not "know" how to fold a protein, so as to make it work. Therefore, it cannot really be considered "the secret of life."

Fourth, the proteins that DNA produces not only may require a special folding process but are also still incomplete, as they may need the presence of additional, non-protein factors that are not under direct DNA control, but must come from the environment. Many proteins, especially those with enzymatic activity, need "helper molecules," called cofactors, so as to perform their biological function. The most common cofactors are metal ions, such as iron, zinc, and copper; other cofactors are vitamins (vitamin C) or are made from vitamins (B-vitamins). Without these cofactors, many DNA products cannot function properly, because DNA may deliver an incomplete protein product.

Where does all of this leave us? The DNA molecule carries "information" in the arrangement of its molecular parts, but the information itself is not composed of physical elements, and the information does not work autonomously. So when some call DNA "the secret of life," we need to stress that this "secret" is more than the biochemistry of DNA—it's also a matter of information processing. This awareness made the cell biologist Barry Commoner turn the mantra "DNA is the secret of life" around into the more accurate claim that "life is the secret of DNA."[136] Without the "machinery" of information processing in the cell, DNA is just a lifeless molecule. It does hold vital information. However, information does not interpret itself.

So the analogy of concepts, such as "information" and "code," borrowed from the informational sciences may be helpful to a certain extent, but it is also deceiving. With the analogy of building a house, we could understand DNA as specifying the materials to be used, but that still leaves the question of where the floor plan comes from. It is definitely not the molecular composition of organisms that determines their form. But if the form is not specified by DNA molecules, where, then, does it come from? Since the same DNA code is present in each part of the body, what tells it which part to develop and where to deploy it?

To say that the information process is guided by DNA can be deceiving in another way. As John Haldane remarks, DNA

[136] Barry Commoner, "Roles of Deoxyribonucleic Acid in Inheritance," *Nature* 202 (1964): 960–968.

may be similar to a "blueprint" in one respect—in that it guides the development of the organism—but the obvious disparity is that the blueprint does not enter into the structure of the house, whereas the genes which contain the DNA, and whose sequence guides the development of a growing organism, remain within—as parts of the embryo, fetus, infant, adolescent, and so on throughout life.[137]

So there must be more to DNA than what we just said. As a matter of fact, genes come in at least two main forms, expressive and regulatory (although some perform both functions). The expressive genes are those that create the cell proteins needed for structure and metabolism, as mentioned earlier. But before a gene can be transcribed into RNA and translated into a certain protein, it may have to be "switched on," which is a function of the genetic regulatory system. Without on- and off-switches, the genetic code would be uncontrollable. It is actually surprising just how little of the genome is used to code for protein—a measly 1.5 percent of the total. That makes DNA even more awesome.

Yet DNA is certainly not the "secret of life"—there is much more to it. From here on, it would be only a small step to claim that it is God who is the real secret of life and of DNA.

DNA as the language of God

If the information stored in the DNA code doesn't come from its carrier, could it perhaps come from God? Could the DNA code ever account for its own origin? No, that's very unlikely—as unlikely as the universe's accounting for its own origin. Nothing can account for its own origin.

[137] John Haldane, *Reasonable Faith* (New York: Routledge, 2010), 140.

Interestingly enough, Francis Collins—the former longtime leader of the Human Genome Project and currently the director of the National Institutes of Health—explicitly speaks about DNA as "the language of God" and about his famous human genome project as "laying open the pages of this most powerful textbook."[138] DNA is the "language" in which God created life. Our human DNA is like a "text" that finds its origin in God, the First Cause of everything in this universe. The "text" of human DNA has turned out to be three billion letters long, written in a four-letter code. This goes almost beyond our imagination; a live reading of that code, day and night, at a rate of one letter per second, would take someone thirty-one years.

Collins is right: God speaks through everything there is, and that includes DNA. Just as mathematics could be called the language of God by Galileo,[139] so could DNA be considered God's language as well. DNA may appear to us to be driven by chance, but the outcome is specified entirely in the Mind of God. DNA can contain information only because it is a reflection of the information found in God's Mind. That's why God can also be found in DNA. One could easily join Francis Collins when he exclaims, "How deeply satisfying is the digital elegance of DNA"[140]—comparable to the rational elegance of the universe and its laws of nature, as we discovered earlier. There is certainly an "elegance" to DNA.

We haven't answered the question yet as to what makes DNA molecules start replicating themselves and transferring their information to proteins. Of course, we can come up with causes that trigger this process—causes such as enzymes and the like. However, that starts an endless process of searching for causes behind

[138] Collins, *Language of God*, 111.
[139] Galileo, *Il saggiatore*, 178.
[140] Collins, *Language of God*, 107.

causes, which goes on into infinity—basically without explaining anything. The only cause to stop this infinite regress is the First Cause, God. Only God is the Cause where "the buck stops." DNA could not do anything, could not even exist, if the First Cause, God, did not make it exist to become a secondary cause of its own—a carrier of information. All the steps in the chain from DNA to RNA to proteins are secondary causes that need a First Cause that lets them be causes of their own—otherwise they could not even act as causes in a chain of cause-and-effect events.

Just as God has implemented the law of gravitation to make planets orbit the sun and prevent us from falling off the earth, so has God put into action the laws and mechanisms of genetics to have us make new generations. Instead of making each one of us from scratch, God gave us genes with DNA, so we can pass on to the next generation the information of how to make a new human body. Had anyone from among us come up with this idea, we would call it ingenious, but when it's God's "invention," we shrug it off and accept it as the most normal thing on earth. But is it really?

To put all of this in a nutshell, we are more than DNA—not "*only*" DNA, but "*also*" DNA. As a matter of fact, we are God's "invention," made from the "dust of the earth."[141] And yet we are more than the "dust" of atoms and molecules. Using the distinction we made earlier—creating versus producing—we might perhaps say now that the DNA code produced us, but God created us. God uses the DNA code just as He uses the laws of nature to make things happen in this world. Our DNA has come to us through our ancestors, but it couldn't exist if it had not come from God. Children come *through* the laws of genetics and the DNA of their parents, but ultimately they come *from* God.

[141] See Gen. 2:7.

7

How Evolution Points to God

Whether you believe in evolution or not is a decision only you can make. If you don't believe in evolution, you may consider this chapter irrelevant. But even then, you will often find yourself in discussion with people who do believe in evolution. So it is an important issue in the current age. You may want to at least find out that belief in evolution does not block belief in God.

Those who do accept evolution need to know that it is only possible on at least two conditions. First, there must be genetic diversity or variability for evolution to work with. The driving force of genetic diversity is random mutation, mainly in the DNA code. "Randomness" is a key word here.

Second, there is a process of natural selection, which then works on this genetic diversity and gives some variants of a gene a better chance than others to be transferred to the next generation. This selective filter may change the genetic constitution of the future generation. "Selection" is the key word here.

Is evolution steered by randomness?

The idea that evolution is steered by randomness is based on the fact that mutations are random. However, how are we to understand randomness? The word "random" is a scientific, actually a statistical or stochastic, concept. When people toss a coin, there is randomness involved because the outcome is independent of what the coin-tosser would like to see, and it is independent of previous and future tosses. Yet the outcome is predictable in terms of statistical probabilities.

When it comes to mutations, the term "random" has a more specific, technical meaning. Mutations are random in several ways: they are "spontaneous" in the sense that they just pop up; they are "unpredictable" as to where they will strike in the DNA; they are "arbitrary," because they hit good and bad spots alike, so to speak; they are "aimless," because they occur without any clear connection to immediate or future needs of the organism. When biologists say that mutations are "statistically random"—they mean mutations just happen one way or the other, because chance has no "favorites," no "memory," and no "foresight."

This sounds like a rather wild process. However, randomness cannot steer evolution in just any direction. Certain directions are not possible, because randomness is kept in tow by the laws of nature. One of the main restraints put on mutations is the so-called Periodic System. This system regulates which combinations of atoms and molecules are possible, and which reactions they may undergo.

Atoms and molecules have "built-in" affinities to each other that allow only for certain combinations and reactions. Helium and neon, for instance, are the two lightest elements in a chemical group known as "noble" gases. These elements usually do not take part in chemical reactions because all the atoms in the group—helium, neon, argon, krypton, xenon, and radon—have outer electron shells that are fully filled, with neither any spare electrons to donate to other atoms, nor any electron deficits that could be filled by electrons donated from other elements.

The Periodic System determines and explains all of this. The electron configuration or organization of electrons orbiting the nucleus of neutral atoms shows a very specific, recurring pattern. Electrons float around the nucleus in different shells or energy levels. The lower the energy level of the electron, the closer it is to the nucleus.

A principal energy level may contain up to $2n^2$ electrons, with *n* being the number of each level. The first energy level can contain $2(1)^2$ or two electrons; the second can contain up to $2(2)^2$ or eight electrons; the third can contain up to $2(3)^2$ or eighteen electrons, and so on. As the atomic number increases, electrons progressively fill these shells. The shells of lower energy are filled in with electrons first, and only then are the higher energy levels filled. This applies to all ninety-two naturally occurring elements.

Electrons obviously play an important role in all of this. So-called "free radicals," for instance, are atoms, molecules, or ions with unpaired electrons. With some exceptions, these unpaired electrons cause radicals to be highly reactive; they are prone to losing or picking up an electron, so that all electrons in the atom or molecule will be paired. Take, for instance, the free radical CO_3. In nature we find CO (carbon monoxide), CO_2 (carbon dioxide), but no CO_3 (carbon trioxide). Yes, carbon trioxide does exist, but it is so unstable that it degrades into CO_2 and H_2O—carbon dioxide and water—with a lifetime much shorter than one minute.[142] To make a long story short, chemical reactions, including mutations in DNA, are not as completely random as they may appear. There are regularities and constraints.

What we may conclude from this is that the world of atoms is not a chaotic world, but has the elegant structure of the Periodic System. It's a structure that was discovered by the Russian chemist

[142] W. B. DeMore and C. W. Jacobsen, "Formation of Carbon Trioxide in the Photolysis of Ozone in Liquid Carbon Dioxide," *Journal of Physical Chemistry* 73, no. 9 (1969): 2935–2938.

Dmitri Mendeleev around 1870. But he didn't invent it: he discovered it. Had he designed it on his own, we would say he was an ingenious designer. Why wouldn't we say the same about the design of the atomic elements? They have so much harmony and elegance that it's hard *not* to see the Periodic System as pointing to a Cosmic Designer. What else could explain such a beautiful system?

Yet the phenomenon of randomness remains standing. It definitely plays a role in evolution and may, therefore, create the impression that randomness has taken over the role of God in evolution. How can a random outcome ever be connected to God, for randomness and God can hardly go together? As a matter of fact, many scientists think it has indeed replaced God who, if He exists, is only "playing dice." Do they have a point?

Not really. Replacing the role of God with the role of randomness basically means replacing God with randomness as the First Cause, thus making randomness the ultimate explanation of all other causes. Why can this not be true? If randomness is the basis for *change* in the universe, it must be a secondary cause and cannot be itself a first cause, because chance events occur *within* nature, and therefore must be a secondary cause.

The real First Cause can use random secondary causes similar to the way we can use dice to determine who wins a game. So when the outcome of a mutation is considered random, that does not tell us that God has nothing to do with it. Even random events such as mutations are only possible because they depend on a First Cause and were created by a First Cause, God—otherwise, they could not even exist.

This suggests that God and randomness are not in conflict with each other and do not exclude each other. The role of randomness concerns the relationship between secondary causes, whereas the "role" of God is about the relationship of secondary causes to the First Cause. Therefore, anything that seems to be random from a scientific point of view may very well be related to God at the same time. Stephen Barr makes a similar distinction when he says,

> No measurement, observation, or mathematical analysis can test whether or not God planned a development like a genetic mutation. What apparatus would one employ? Being "unplanned by God" is simply not a concept that fits within empirical science. Being "statistically random," on the other hand, is, because it can be tested for.[143]

The idea that randomness can play a role in God's relationship to the world is a rather common idea among Catholic thinkers. Take, for instance, St. Thomas Aquinas who once said, "God causes chance and random events to be the chance and random events which they are, just as he causes the free acts of human beings to be free acts."[144] Yet he cautioned against dismissing divine providence: "Whoever believes that everything is a matter of chance, does not believe that God exists."[145] Or take St. Padre Pio, who was apt to say in various ways that it is God who arranges the coincidences. Once he asked a man who claimed such-and-such event had happened by chance, "And who, do you suppose, arranged the chances?"[146] Science has no answer to this question—not even the answer "nobody did." Anything that seems to be random from a scientific point of view may very well be included in God's eternal plan. Cardinal

[143] Stephen M. Barr, "Chance, by Design," *First Things* (December 2012): 26.

[144] Quoted in William E. Carroll, "Evolution, Creation, and the Catholic Church," lecture at Williams College, October 19, 2006.

[145] Aquinas, *De Symbolo Apostolorum* 4, 33.

[146] Quoted by Mark Brumley, "The Mystery of Human Origins: Which theories are compatible with Catholic faith?," *The Catholic Answer*, January–February 2005.

John Henry Newman wrote in an 1868 letter, "I do not [see] that 'the accidental evolution of organic beings' is inconsistent with divine design — it is accidental to us, not to God."[147]

However, there is still a potential problem with mutations. Some of them have serious harmful consequences — they are "failed trials." Can such genetic errors be connected with God? Yes, they can: God can be in genetic errors in the same way He can be in mutations and randomness. But that doesn't mean genetic errors are directly caused by God, let alone explicitly willed by Him. Yes, they are God's doing, but not in the sense of being directly caused or willed by God. Those errors are secondary causes of genetic defects and diseases, which are themselves caused by other secondary causes. However, what remains true is that secondary causes cannot exist without the First Cause, God. God is the Maker and Master of *all* there is.

As we said before about mutations and randomness, we must distinguish between "God-talk" and "science-talk." Genetic errors are "science-talk" — they are about the relationship of secondary causes to each other, not the relationship of secondary causes to the First Cause, God. Genetic errors are caused by radiation, mutation, mutagens, and the like. We cannot blame God for them. Genetic errors are not God's doing in a direct sense — they are not caused by God directly but by other secondary causes. Most of all, they happen due to the laws of nature, which are part of God's creation as well.

Because of this, we do not have to wonder about God's will every time a genetic error occurs any more than we have to wonder about God's will every time a stone falls to the ground, even if it strikes us on the head. All that happens in this world must be seen in this

[147] John Henry Newman, "Letter to J. Walker of Scarborough, May 22, 1868," in *The Letters and Diaries of John Henry Newman* (Oxford: Clarendon Press, 1973).

context: as earthquakes and floods are a necessary consequence of the structure of the physical world, so are viruses and genetic errors a necessary consequence of the fabric of the biological world. Such explanations cause events like these to happen without making God directly responsible. As Michael Augros puts it, "If secondary causes, unlike the Primary Cause [First Cause], are not infallible, if they are defectible, then we might well blame them, rather than the first cause, for any flaws we find (or think we find)."[148]

Obviously, the forces of nature and the laws of nature can cause results that are lovely as well as cruel and beautiful as well as ugly. Volcanos can create beautiful islands and mountains but also devastating destruction. The power of growth makes flowers and babies develop into something beautiful, but it also makes tumors get bigger and bigger. Weather patterns may be the cause of a gentle breeze as well as a destructive tornado. DNA helps us to be who we are and to do what we do, but it can also go against us. Mutations gave us the diversity of life we see around us, but they can also destruct what they had once produced. Obviously, nature is a combination of forces following laws that can have both constructive and destructive consequences.

Has natural selection replaced God?

Randomness in evolution basically creates new "trials" that have to be tested next in the "laboratory" of natural selection. Randomness creates genetic diversity, but it is natural selection that filters this diversity. Put in an image, natural selection is only the editor, not the author, of evolutionary change.

The concept of natural selection was first introduced by Charles Darwin. Even Darwin himself always felt uneasy about his term

[148] Ibid.

"natural selection," because it leads almost automatically to an obvious question: "Selection by whom or by what?" Indeed, the implication of a "selecting agent" looms large—something Darwin wanted to avoid by saying he had as much right to use metaphorical language as physicists do. In his own words, "Who objects to an author speaking of the attraction of gravity as ruling the movements of the planets? Everyone knows what is meant and is implied by such metaphorical expressions."[149] Yet, at times, he does refer to "nature" as some kind of feminine deity, being the agent of selection.

Eventually, though, his colleague Alfred Wallace convinced Darwin to replace the term "natural selection" with Herbert Spencer's notion of "survival of the fittest"[150] in the fifth edition of his book. Nevertheless, the question "Selection by whom or by what?" remains pressing. Although Darwin wanted to avoid this question, it does in fact have an answer: selection is done through the laws of nature implemented by a Cosmic Lawgiver.

Nevertheless, natural selection is often seen as a replacement for God. It is not quite clear whether Darwin himself thought it was. He often called himself an agnostic rather than an atheist.[151] Perhaps he saw natural selection as a new law of nature designed by God. But many of his followers saw it as a substitute for God. For them, it meant the end of their belief in Creation by God, whose role in evolution had been replaced by natural selection. However, the truth of the matter is that natural selection is hardly conceivable without God. Let's see why.

Natural selection works by definition on organisms with "successful" features. A biological design can be successful only if it is

[149] Charles Darwin, *The Origin of Species*, 5th ed., chap. 4, 93.

[150] Herbert Spencer, *Principles of Biology*, vol. 1 (London: Williams and Norgate, 1864), 444,

[151] See, for example, his letter to John Fordyce, dated May 7, 1879.

in accordance with the laws of nature. Without these natural laws, biological designs could not work at all. A heart could not pump blood, nor could a fish swim, if it did not follow hydrodynamic laws. Without following aerodynamic laws, a bird's wing could not allow flight. Natural selection relies on the laws of nature, which, as we found out earlier, point to a Divine Lawgiver. Natural selection, therefore, must also point to God.

Therefore, the "fittest" are not defined by their survival—that would make for a tautology ("the fittest are the ones who survive")—but by their design. Consequently, biological fitness is not an outcome of natural selection, but a condition for it. What is it that makes organisms "fit"? Or, put differently, what is it that carries certain biological designs or patterns to the next generation? The answer would go like this. The universe has an overall set of restraints harnessing individual designs and making them "fit" or "successful" to a certain degree. For example, the fittest or most successful bird is the one following the laws of aerodynamics the closest. And the fittest or most successful fish is the one following the hydrodynamic laws the closest.

Given the way the universe is designed, some biological designs are better than others by having a better outcome—a better "fit," so to speak, which makes them more successful in reproduction and survival. They must have "something" in their biological design that carried them through the "filter" of natural selection. Natural selection can select only those specific biological designs that are in accordance with the laws of the cosmic design. Natural selection does not *explain* a "fit" but *uses* a "fit" in order to select what fits best. Being "fit" comes from the way the universe is designed. And where does the design of the universe, with its physical constants and laws of nature, come from? It is hard not to point to God for an answer to that question.

Does this mean that the course of evolution had to be the way it is? We don't fully know, certainly not in purely scientific terms. Perhaps the best we can say is that the evolutionary road to humanity is a process that meanders like a river. On the one hand, it follows a path that seems coincidental and random. On the other, in spite of its winding flow, it also moves in a specific direction, steered by natural selection, which is like a path of least resistance. Just as a river follows a path of least resistance according to the topographic design of the landscape, so does the "stream" of evolution follow a path somehow regulated by the design of our universe.

To put it in a nutshell, evolution follows the "path of least resistance" in the "landscape" of the cosmic design. So it does not just flow in a purely random manner. Somehow, the cosmic design creates the "bed" in which the stream of evolution meanders. Amazingly enough, the flow of evolution found its culminating destination in the rise of humanity.

No evolution without God

It was in 2005 that thirty-eight (!) Nobel laureates issued an open letter to the Kansas Board of Education in defense of evolutionary theory, with the words, "Evolution is understood to be the result of an unguided and unplanned process of random variation and natural selection." It's stunning how they could say this while claiming to be backed by their scientific expertise. As Stephen Barr rightly remarks, "When biologists start making statements about processes being unsupervised, undirected, unguided, and unplanned, they are not speaking scientifically."[152] The same could be said about those thirty-eight Nobel laureates who made their blatantly unscientific statements in spite of all the arguments against their extravagant claims.

[152] Barr, "Chance, by Design," 26.

There is no way scientists can speak in terms of "unsupervised, undirected, unguided, and unplanned." Those concepts are not part of scientific terminology. They have actually been specifically removed from scientific vocabulary. The concept of "purpose," for instance, was taken out of astronomy by Nicolas Copernicus, out of physics by Isaac Newton, and out of biology by Charles Darwin. Once such terms have been eliminated from science, they can no longer be used, let alone be explained, by science, as they are no longer part of scientific vocabulary.

At the same time, they cannot possibly be removed entirely from human discourse. When scientists removed "purposes" from scientific discourse, they removed them as secondary causes, but they left their reference to the First Cause untouched. So they did not make purposes disappear completely; they just moved them from inside to outside the scientific domain—from the context of secondary causes to the context of the First Cause. But that doesn't make them vanish entirely.

Those who keep denying the existence of any purposes at all need to ask themselves some pertinent philosophical questions. If there is no purpose in the universe, tout court, how then were we ever to know there is no such thing as a purpose? As C. S. Lewis put it, "If there were no light in the Universe and therefore no creatures with eyes, we would never know it was dark."[153] Besides, we should ask those who deny the existence of purposes what the purpose is of arguing that there is no purpose in life. As a matter of fact, denying that there are purposes in life defeats its own claim. The reason is simple: if it's your purpose to remove all purposes from life, then you are also wiping out your own purpose of trying to do so.

Scientists have no right and no good reason in their role as scientists to decree that we as human beings are unintended, unplanned,

[153] C. S. Lewis, *Mere Christianity* (San Francisco: Harper, 2001), 46.

unguided, fortuitous creatures, or mere products of a blind and purposeless fate. Science has nothing to say about such issues. Evolutionary theory cannot even explain why there is evolution to begin with. Can we still talk about evolution in terms of purposes and the like? Could evolution still be seen as steered by the purposes God has in mind? I don't see why not. The case could even be made that evolution cannot be understood without any reference to God. Let me explain why.

If we assume or declare that the theory of evolution completely explains the course of evolution from simple organisms to complex human beings, then we get into serious trouble, for now the question arises as to where that theory itself comes from. The answer must be that the theory of evolution originated in the mind of Charles Darwin, and now still resides in the minds of many biologists. But then we must question how trustworthy this theory is, given the fact that this very theory must be a product of evolution, too. Doesn't that outcome defeat its own claims? Doesn't that lead to self-destruction?

Interestingly enough, even Charles Darwin himself vaguely acknowledged this problem when he said in his autobiography, "But then with me the horrid doubt always arises whether the convictions of man's mind, which has been developed from the mind of the lower animals, are of any value or at all trustworthy. Would anyone trust in the convictions of a monkey's mind, if there are any convictions in such a mind?"[154] The theory of natural selection gives him reason to wonder whether, as he puts it, "The mind of man, which has, as I fully believe, been developed from a mind as low as that possessed by the lowest animal, [can] be trusted when it draws such grand conclusions."[155] Certainly, Darwin had good reason to worry about the trustworthiness of those "grand conclusions."

[154] Darwin's letter to W. Graham, 1881.

[155] Charles Darwin, *The Autobiography of Charles Darwin* (Cambridge, UK: Icon Books, 2003), 149.

Ironically, when Darwin said this, he questioned only the "grand conclusions" regarding belief in the existence of God. What Darwin did not seem to realize is that the theory of evolution is another "grand conclusion." So he should have the same "horrid doubt" about the trustworthiness of his own theory of evolution. Apparently, he didn't, whatever his reason may have been. He must have tacitly assumed that his mind was able to find the truth, regardless of its evolutionary history. But if so, then his mind cannot be the mere product of evolution.

If we explain the existence of the human mind exclusively in terms of evolution, we have created a "boomerang" that will eventually hit whoever launched it. The problem we have here is that evolution can neither create nor explain the existence of the human mind, but must assume it. If the human mind were really the mere product of evolution, so would the science of evolution—hence, nothing we claim to know could be trusted. All scientific discoveries would be mere illusions concocted by a neural network under the shaky direction of genes shaped by natural selection.

So we must come to the conclusion that evolutionary theory only reveals us one of the many aspects of the universe, its evolutionary aspect. But there are many other aspects, one of them being the aspect of the world's relationship to God. So evolutionary theory on its own is missing an essential and crucial part of the entire puzzle. That's what the great English poet Alexander Pope noticed in the mid-eighteenth century about the atheists of his day: they "see Nature in some partial narrow shape, / And let the Author of the Whole escape."[156]

Evolution and evolutionary theory cannot give us the entire picture; they leave out an essential part, God. But we cannot let the "Author of the Whole" escape, for our knowledge would thus

[156] Alexander Pope, *The Dunciad*, bk. 4, 455–456.

collapse. The *Catechism of the Catholic Church* puts all of this together as follows: the world "is not the product of any necessity whatever, nor of blind fate or chance."[157] In his first homily as pontiff, in 2005, Pope Benedict XVI insisted, "We are not some casual and meaningless product of evolution. Each of us is the result of a thought of God. Each of us is willed, each of us is loved, each of us is necessary."[158] This remains true, even in a world seemingly laid bare by science. That's why randomness and divine providence are not enemies. When scientists focus only on the scientific details, they tend to lose sight of the larger picture.

In a nutshell, all organisms may have come *through* a process of evolution, but ultimately they must come *from* Creation, or else they couldn't be here at all. Whereas evolution tells us that everything comes from something else, Creation tells us that everything would be nothing if it didn't come from God. An evolutionary world depends on God as much as would a world without evolution. Evolution follows the laws laid out in Creation, but Creation is the origin and destination of evolution. Humanity may have come here through evolution, but ultimately it comes from God.

[157] CCC 295.

[158] Benedict XVI, Homily at the Mass for the inauguration of his pontificate (April 24, 2005).

8

How Neuroscience Points to God

Neuroscience studies the structure and function of the nervous system—more specifically, of the brain. The Greek physician Hippocrates surmised already that the brain is not only involved with sensation—since most of its organs (such as eyes, ears, and the tongue) are located in the head near the brain—but is also the seat of intelligence. Plato, too, speculated that the brain was the seat of the rational part of the soul. The Roman physician Galen, a follower of Hippocrates and physician to Roman gladiators, had noticed that his patients lost their mental faculties following brain damage.

Nowadays, we know much more about the brain and its neural network, but how all of this is connected with our mental faculties is still very much subject of debate. If all our thinking is merely a brain issue, then there is no reason for further discussion. But if our faculty of thinking is more than a brain issue, there may be much more that is pointing to God. Just as the laws of evolution point to God, so do the laws of the neural network ultimately point to God.

A matter of thoughts

Earlier in this book, we discussed the doctrine of materialism and the role it seems to play in the minds of many scientists. However, if materialism were really true, then what are we to make of *thoughts*? How can thoughts be material? Of course, we can record our thoughts with a dictaphone, which makes them material, and we can write our thoughts down on paper, which "materializes" them in another way. But what about the thoughts themselves? How can thoughts be material in and of themselves? That's a rather distressing question.

This momentous, skeptical question won't discourage materialists, though—and many neuroscientists are materialists. They will quickly respond that thoughts are, in fact, material entities as well. Thoughts, so they say, are just a certain pattern of electrical impulses in the neurons of our brains.

But then, immediately, another serious problem arises. Unlike neurons, thoughts do not have any material characteristics such as length, width, height, and weight. Thoughts can be true or false, right or wrong, but never tall or short, heavy or light—they have no mass, no size, no color. We can think about sizes and colors of things, but the thoughts themselves do not possess these qualities. As Alvin Plantinga puts it, "It's a little like trying to understand what it would be for the number seven, e.g., to weigh five pounds.... A number just isn't the sort of thing that can have weight."[159]

So to evaluate the outcome of neural states as either true or false, we need something that is *not* neural, for the simple reason that a pattern of electrical nerve impulses can't be true or false. As John C. Polkinghorne puts it, "Neural events simply happen, and that is that."[160] So there is no way even to think in terms of true or false, for as Stephen Barr notes, "One pattern of nerve impulses

[159] Alvin Plantinga, "Against Materialism," *Faith and Philosophy* 23 (January 2006): 14.

[160] Polkinghorne, *Science and Creation*, 36.

cannot be truer or less true than another pattern, any more than a toothache can be truer or less true than another toothache."[161]

Unlike brain processes, which are subject to physical or neural causation, thoughts are subject to mental causation based on reason and intellect, on laws of logic and mathematics. The thought "one plus one," for instance, does not physically cause the answer "two"—if it did, we could have nicely skipped some classes in school. We may have been trained to answer that way, but the correct answer is based on math, not on training. In other words, the brain does not secrete thoughts the way the pancreas secretes hormones. Even brain scans can pick up "brain waves," as some like to call them, but never thoughts, for their immateriality prevents them from showing up on pictures and scans.

Apparently, there is something very peculiar about thoughts: they have *content*—they are *about* something, specifically about something beyond themselves. But how could an assemblage of neurons—a group of material objects firing away—have any *content*? To use an analogy, anything that shows up on a computer monitor remains just an "empty" collection of dots on the screen that do not point beyond themselves until some kind of human interpretation gives sense and meaning to the display on the screen by interpreting it as being about something else. In a similar way, a thought may have a material substrate in a physical network, such as the brain or a computer, but this substrate acts only as a physical "carrier" for something immaterial—thoughts, that is—in the same way newspapers carry thoughts of journalists. To think otherwise is to say that Shakespeare's thoughts are nothing but ink marks on paper.

This outcome has serious consequences for materialism itself. If materialists want to claim that materialism is true, then they should realize that their very *thoughts* about materialism must be more than

[161] Barr, *Modern Physics*, 197.

a certain pattern of electrical activity in their brain cells. If thoughts were really material, based on neurons, impulses, and neurotransmitters, then they would be just as fragile as these materials. Besides, reducing thoughts to a creation of neurons in the brain obscures the fact that the word "neuron" is itself an abstract, immaterial entity in our thoughts. Such a claim starts a vicious circle.

Stephen Barr, for one, shows us the vicious circle as follows: "The very theory which says that theories are neurons firing is itself naught but neurons firing."[162] If materialists claim materialism is true, then they should realize this very thought must be more than a certain pattern of electrical activity in their brain cells; otherwise, materialism works like a boomerang that undermines its own foundation and makes its claims self-destructive. The statement that there are only material things is as fragile as the "material" that supposedly generated this worldview.

Yet there is a rather common misconception that materialism is backed by science. Although many scientists are materialists, materialism is not science, but merely a philosophical opinion. The most science can offer us is the observation that many things in this universe are material and can be quantified, measured, counted, and dissected. However, talking about "many things" does not entitle us to talk about "all things"—there is no way we could conclusively reason from "many" to "all." After seeing many white swans, for instance, one cannot safely conclude *all* swans are white. After seeing many material things, one cannot safely conclude that *all* things on earth are material. Such a conclusion is not logically justified—there is no way of knowing. One cannot even defend such arguments by saying they have worked so many times in the past, for that would be another example of moving from "many" to "all." This would take us on a never-ending search for proof.

[162] Barr, *Modern Physics*, 196.

We can find another reason why science on its own can never prove that matter is all there is in the fact that science first limits itself exclusively to material things and then claims to show us that there is actually nothing but material things. Science excludes immaterial entities from our discourse ahead of time, and then many scientists "conclude" from this there are only material entities. That is basically an example of circular reasoning: it begins with what it is trying to end with—that's how we keep circling around.

However, if matter were indeed all there is, then one should wonder what materialism itself is—another piece of matter? Clearly not. So there must be more than matter. This leaves room for other nonmaterial things, such as logic, mathematics, philosophy, morality, and ultimately religious faith. Materialists, in contrast, have a very limited outlook on things. When they come across a phenomenon that is hard to account for in terms of materialism, they often end up just denying its very existence.

Ironically, despite all of the above, materialism still has quite a "spiritual" appeal, for some enigmatic reason. It allows for only one way of looking at the world. It is monopolistic by nature, with no tolerance for competitors. But there should instead always be room for other views and perspectives on the world. Nothing entitles us to ignore that there is more to life than the material entities of molecules, neurons, and genes. If materialism is true, then immaterial entities such as thoughts cannot exist by themselves. And yet they do. In other words, there is so much in life that the thermometers and Geiger-counters of materialism cannot possibly capture—things such as thoughts, concepts, values, beliefs, laws, experiences, hopes, dreams, and ideals. There is no way materialism can deal with these—other than denying them, but then it must deny itself as well.

So we must come to the conclusion that the doctrine of materialism would be at best a dogmatic conviction, certainly not a scientific discovery or a conclusion of the empirical sciences. It is actually based on a circular argument: materialism is true, because materialism *must* be true. It seems safe to say that materialism is based on assumptions rather than proofs. In other words, it *assumes*, in advance, that nonmaterial objects cannot exist. But that assumption actually turns out to be self-destructive—it undermines its own nonmaterial assumption. As Stephen Barr puts it, "Just as the astrologer believes that his life is controlled by the orbits of the planets, the materialist believes that his own actions and thoughts are controlled by the orbits of the electrons in his brain."[163]

Michael Augros brings this argument to a close: "Matter itself is a product, receiving its very existence from the action of something before it."[164] So God remains standing as the only First Cause; it is God who brings matter into existence and sustains it. Matter cannot do so on its own. That's why matter cannot be a first cause, regardless of how "fundamental" its role seems to be in life. Nothing, not even matter, can just pop itself into existence; as we said earlier, it must have a cause, because it does not and cannot have the power to make itself exist. Besides, if matter were a first cause, it would be necessary instead of contingent. This would mean that everything about matter could be deduced by pure thinking, without any observations or experiments—which is absurd too.

Therefore, matter needs an explanation beyond itself, for neither can it explain its own existence nor can it replace God. The Nobel laureate and neurophysiologist John C. Eccles quite accurately described materialism as "a religious belief held by dogmatic materialists ... who often confuse their religion with their science."[165]

[163] Barr, *Modern Physics*,17.

[164] Augros, *Who Designed the Designer?*, 63.

[165] John C. Eccles, *The Wonder of Being Human: Our Brain and Our Mind* (San Francisco: Shambala, 1985), 24.

Brain versus mind

If you want to know what the concept of the brain stands for, study neuroscience. But what does the concept of the mind represent? It seems to be a rather vague idea, one that is hard to nail down. Let me try this: the human mind is the intellectual part of the human soul. That seems to make things worse, for the soul is even more intangible than the mind.

No wonder, then, that many scientists have thrown out the concept of mind by reducing the mind to the brain—to something much more "tangible." They have become what I like to call "brain-mind equalizers." So what some take to be uniquely human faculties—language, rationality, morality, self-awareness, and religion—supposedly don't exist, except as part of the brain.

We are told, too, that mutations changed the brain during the rise of humanity. As a consequence, all our presumably "unique" human faculties can be resolved by reducing them to mutations that altered the brain to the way we know it now. In other words, if the mind is identical to the brain, and if mental issues are nothing but neural issues, then the discussion is closed, and everything that happened during the rise of humanity can be entirely explained by genetics and neuroscience.

However, if the mind is not the same as the brain, the discussion is far from over. We have reached here a critical point in this debate. The pivotal question is this: Is the mind identical to the brain? Many scientists, including neuroscientists, would say it is—they are happy to join the brain-mind equalizers. But what would happen if they were wrong?

Well, there are several reasons why it may not be possible to equate or reduce *mental* phenomena to *neural* phenomena. Some of these reasons are scientific, some empirical, some epistemological, and some just common sense. To find out what they are worth, let's study them in more detail.

Reason 1: One of the pioneers in neurosurgery, Wilder Penfield, made a compelling case about the difference between mental events and neural events when, during open-brain surgery, he asked one of his patients to try to resist the movement of the his left arm, which Penfield was about to make move by stimulating the motor cortex in the right hemisphere of the patient's brain. The patient grabbed his left arm with his right hand in order to restrict the movement Penfield was inducing. As Penfield described this, "Behind the brain action of one hemisphere was the patient's mind. Behind the action of the other hemisphere was the electrode."[166] In other words, one action had a physical, neural cause, whereas the other action had a non-neural, mental cause. Therefore, he concluded the physical cause and the mental cause had a different origin and were of a different nature.

From this follows, as the neurologist Viktor Frankl put it, that while the brain does condition the mind, it does not give rise to it. John Eccles concluded from experiments such as Penfield's, "Voluntary movements can be freely initiated independently of any determining influences within the neuronal machinery of the brain itself."[167] This observation seems to call for the existence of a mind in addition to the brain. The cognitive scientist Jerry Fodor put it most vividly and dramatically: "If it isn't literally true that my wanting is causally responsible for my reaching, and my itching is causally responsible for my scratching, and my believing is causally responsible for my saying ... if none of that is literally

[166] Wilder Penfield, in the Control of the Mind Symposium, held at the University of California Medical Center, San Francisco, 1961. Quoted in Arthur Koestler, *Ghost in the Machine* (London: Hutchinson Publishing Group, 1967), 203–204.

[167] Sir John Eccles and Karl R. Popper, *The Self and Its Brain* (New York: Springer Verlag International, 1977), 294.

true, then practically everything I believe about anything is false and it's the end of the world."[168]

Reason 2: When neuroscientists claim that certain mental phenomena are associated with certain neural phenomena, they cannot conclude from this that these mental phenomena were *caused* by neural phenomena. Correlation doesn't automatically lead to causation, as we discussed earlier. There is a statistical correlation between the sales of ice cream and the sales of sunglasses, but that does not mean one causes the other. We all know that the rooster's crow does not *cause* the sun to rise, and that rotating windmills do not *cause* wind. In a similar way, the fact that regions light up during functional magnetic resonance imaging (fMRI)[169] does not explain whether this lit-up state indicates they are causing a certain mental state, or just reflecting it. Correlation is not always causation.

Nevertheless, brain-mind equalizers use these so-called localization studies to make their case. When certain kinds of mental activity occur, certain parts of the brain do display increased blood flow and increased electrical activity. This correlation makes them conclude it is the brain that causes the mental activity.

However, Alvin Plantinga, for one, does not buy this conclusion: "There are many activities that stand in that same or similar relation to the brain. Consider walking, or running, or ... moving your fingers: for each of these activities too there is a part of your brain related to it in such a way that when you engage in that activity, there is increased blood flow in that part.... Who would conclude that your fingers' moving is really an activity of your

[168] Jerry Fodor, *A Theory of Content and Other Essays* (Cambridge, MA: Bradford Books / MIT Press, 1990), 156.

[169] fMRI measures brain activity by detecting associated changes in blood flow.

brain and not of your fingers? Your fingers' moving is dependent on appropriate brain activity; it hardly follows that their moving just is an activity of your brain."[170] Then, he comes to the conclusion that the mind's *dependence* on the brain is one thing, but *identity* between mind and brain quite another.

It is indeed true that we cannot understand things without using our brains, but it does not follow that our brains are doing the understanding. In other words, brain activity may be a necessary condition for mental activity, but it does not seem to be a sufficient condition. In fact, there are situations where the most intense subjective experiences correlate with a dampening—or even cessation—of brain activity. What comes to mind are cases of near-death experiences or out-of-body experiences. In such cases, neural activity is low or absent, and thus fails not only to be a sufficient condition for mental activity but, perhaps, even to be a necessary condition.

Reason 3: Another peculiarity in this discussion is that something like pain, for instance, can be induced in a physical way, but there is no evidence that experimental stimulation of specific neuronal areas is able to produce a specific mental state, let alone a specific thought. The presumed jump from physical matter to "thinking matter" appears to be enormous.

As the late philosopher Mortimer J. Adler emphasized, there is a clear difference between perceptual and conceptual thought.[171] Thinking in concepts such as "triangle" or "gravity" requires universality, whereas sensory information is about particulars only. The fact that the brain is a necessary, but not a sufficient, condition of *conceptual* thought, indicates that an *immaterial* intellect

[170] Plantinga, "Against Materialism," 23.

[171] Mortimer J. Adler, *Intellect: Mind over Matter* (New York: Macmillan, 1990), chap. 4.

is required also in order to provide an adequate explanation of conceptual thinking. We can even conceptualize what we cannot visualize—a circle with four dimensions, for instance—which calls for something mental, not neural. Very often, we do not see what receptor cells and neurons tell us to see, but rather what we wish to see, which is then, too, not a neural but a mental issue.

Nowadays, there is also growing evidence that the neurocircuitry of the brain is not as static and unchangeable as long thought. So-called "brain plasticity" gave Dr. Norman Doidge the idea that the brain can change itself.[172] But that would be sheer magic. Thomas Aquinas would say that whenever something undergoes change, something must be causing that change. As nothing can cause itself to exist, so nothing can cause itself to change. The brain could be changed by the mind, but not by itself, for every change requires a cause. The brain cannot rewire itself; it needs something else that causes it to rewire. This may well be the mind. However, that would be possible only if the mind is not identical to the brain.

Reason 4: The German philosopher Gottfried Leibniz once suggested to picture the brain so much enlarged that one could walk in it as if in a mill.[173] Inside, we would observe movements of various parts, but never anything like a thought. For this reason, he concluded, thoughts must be different from physical and material movements and parts. Nowadays, the mechanical model of cogs and wheels, which Leibniz used, has been replaced by a model of neural and biochemical pathways, but the outcome is still the same. How can an assemblage of neurons—a group of material objects firing away—have any *content*?

[172] Norman Doidge, *The Brain That Changes Itself: Stories of Personal Triumph from the Frontiers of Brain Science* (New York: Viking Press, 2007).

[173] Gottfried Leibniz, "Monadology 17," in *Leibniz Selections*, ed. Philip Weiner (New York: Charles Scribner's Sons, 1951), 536.

Alvin Plantinga describes the problem in all clarity: "What is it for this structured group of neurons, or the event of which they are a part, to be related, for example, to the proposition *Cleveland is a beautiful city* in such a way that the latter is its content? A single neuron (or quark, electron, atom or whatever) presumably isn't a belief and doesn't have content; but how can belief, content, arise from physical interaction among such material entities as neurons?"[174]

Then he says, "Propositions are also mysterious and have wonderful properties: they manage to be about things; they are true or false; they can be believed; they stand in logical relations to each other. How do they manage to do those things? Well, certainly not by way of interaction among material parts."[175] Just as the brain cannot distinguish between legal and illegal narcotics, so is the brain incapable of telling false thoughts apart from true beliefs. In order to evaluate the outcome of neural states as true or false, we need something that is not neural.

Reason 5: It is hard to equate the working of the mind to the working of a machine such as a computer—although the computer is a popular model for the brain nowadays. Computers require a human maker and would still need a human subject to give their informational output some meaning or sense. Without human subjects, computers just cannot "think." This means computers cannot explain the human mind, but they must presume its existence. Computers do not create thoughts, but they may carry thoughts that were created by the mind of a human subject—namely, the programmer of the computer.

Think of a voice-recognition system: it doesn't really understand what it is programmed to "recognize." Computers do only

[174] Plantinga, "Against Materialism," 14.

[175] Ibid., 21.

what we, human beings with a mind, make them do, for we have proven to be champion machine-builders. So the popular slogan "Man versus Machine" is actually very deceiving; it should be "Man versus Man"—man versus the man who designed the machine.

Besides, even though a computer may play chess better than Kasparov or any other champion, it plays the game for the same "reason" a calculator adds or a pump pumps—the reason being that it is a machine designed for that purpose—and not because it "wants" to or is "happy" to do so. Computers, and similar devices, do not have meaning or sense in themselves until a human subject uses them as carriers of information that receives sense and meaning from a human subject. This makes it hard to use the computer analogy to understand and explain the human mind fully, for without the human mind, there would be no computers. So we end up in circular reasoning again.

Reason 6: Denying the existence of the human mind is another example of a self-destructive assertion. If the mind were just the brain, then its thoughts would be as fragile as the molecules they are supposedly based on. It would be sitting on a "swamp of molecules," unable to pull itself up by its bootstraps. Some biologists, for instance, claim that we believe what we believe because what we call "truth" emerges from brains shaped by natural selection. But such claims , again, work like a boomerang—if they are true, they become false. The snake of this claim is eating its own tail, or rather its own head.

In other words, denying the existence of mental activities is itself a mental activity and thus leads to contradiction. Ironically, one cannot deny the mental without affirming it. Scholars such as J. B. S. Haldane and C. S. Lewis have worded this paradox along the following lines: if I believe that my beliefs are the mere product of natural selection, then I have no reason to believe my beliefs are true—therefore, I have no reason to believe that my beliefs are the mere product of natural selection.

So if we are looking for a key to understanding ourselves, it will not be in terms of matter and brain, but of mind and soul. The brain is governed by laws of physics, chemistry, and biology, but thoughts are not. As Stephen Barr pithily puts it, "We do not infer the existence *of* our minds, rather we infer the existence of everything else *with* our minds. To put it another way: the brain does not infer the existence of the mind, the mind infers the existence of the brain."[176]

Reason 7: In order to make any claims, especially in science, we must validate them; otherwise, they are worth nothing. If Watson and Crick, Planck and Einstein, Darwin and Dawkins, or any other scientists, were nothing but their neurons, then their scientific theories would be as fragile as their neurons. That would be detrimental to their claims and to their own status as experts in their respective fields. If our mental activities are only the by-product of neural events, then they could be nothing more than illusions or mere sensations at best. If we say we are nothing but a "pack of neurons," neither this very statement nor we who make it would be worth more than its molecular origin.

All the claims we make are mental activities. A mind is necessary to come up with generalizations and abstractions such as the law of gravity, for instance. Newton's mind was able to see beyond the sensory impression of a falling apple. This must be more than a brain issue. The law of gravity didn't come from Newton's brain, but from his mind. The physical world can never be studied by something purely physical, any more than neurons could ever discover and study themselves—neurons studying neurons! This means that the brain could never study itself either—only the mind can.

[176] Barr, *Modern Physics*, 188.

Put in more general terms, the *knowing* subject must be more than the *known* object, for a mind is needed to understand the brain and a subject to study any object. To explain the mind in terms of physics obscures the fact that one would still need to have a mind first before one could even have physics. While idealists such as George Berkeley reduce reality to the mind and deny the existence of matter, materialists reduce reality to matter and deny the existence of the mind. But on what grounds do they assume reality must be of only one kind of substance? Why can't reality consist of both matter and mind?

Reason 8: We discussed earlier that crucial to information theory is the separation of content from the "vehicle" that transports it. No possible knowledge of the computer's materials can yield any information whatsoever about the actual content of its computations. Yet there are some similarities between the operation of the brain and the operation of a computer: both use a binary code based on ones (1) and zeros (0); neurons either do (1) or do not (0) fire an electric impulse, in the same way as transistors either do (1) or do not (0) conduct an electric current.

But that is where the comparison ends. Whatever is going on in the brain—say, some particular thought—may have a material substrate that works like a binary code, but this substrate acts only as a physical "carrier" for something immaterial. It would not really matter whether this material substrate works with impulses, as in the brain, or with currents, as in a computer, or with letters, as in a book, for the simple reason that this material substrate merely carries something coming from the mind, not the brain. Firing neurons are simply carriers of immaterial thought.

As said earlier, thoughts are *about* something mental, about something beyond themselves. A mere collection of ones and zeros in the neural network is not about anything until the mind gives sense and meaning to the code and interprets it as being *about* something else. Think of what we call a picture: a picture may carry information, but the picture itself is just a collection of tiny dots; it is merely a piece of paper that makes "sense" only when human beings interpret the picture as being *about* something. The same with books: they provide lots of information for bookworms, but to real worms they have only paper to offer. Again, this means that the neural carrier of information cannot be the same as the mental information it carries.

Reason 9: As stated under Reason 3, it does not seem very likely that thoughts can be induced in a physical way. We are not talking here about emotions or feelings, for instance, because those are physical and biological phenomena that can be physically induced by stimulation of certain brain areas or the use of chemicals. Neither are we referring here to memories stored in the brain—including memories of thoughts once produced by the mind—because memories can be physically "stored," similar to the way thoughts can be "stored" on paper.

Thoughts, on the other hand, cannot be produced in a physical manner, neither by chemicals nor by electrodes. If we could change beliefs and convictions with chemicals, presidential candidates would be wise to use that method in their campaigns. If the thought of "two to the power of two" physically produced the thought of "four," we could have skipped much work in school. Thoughts are not the outcome of physical causation. Unlike brain processes, which are subject to physical causation, thoughts are subject to mental causation based on reason and intellect, on laws of logic and mathematics, which are not hardwired in the brain.

In addition, one could argue that the brain is as much or as little responsible for thinking as the hand is for grasping or the leg for walking. Many think that if one makes the brain responsible for thought, then it somehow becomes the principal agent of thought. However, that idea is as dubious as thinking that if one makes the hand responsible for grasping, then it must be the principal agent of grasping. In fact, the hand is not the agent of grasping, but merely an organ or tool of grasping. In a similar way, the mind uses the brain as its organ and tool. The mind needs the brain to function properly, but the brain also needs the mind to function fully. This would be another argument against the brain-mind equalizers.

Reason 10 is based on a thought experiment the philosopher Ludwig Wittgenstein once suggested.[177] Picture yourself watching through a mirror how a neuroscientist is studying your opened skull for brain waves. It can be stated that the scientist is observing just one thing, outer brain activities, whereas you, the "brain-owner," are actually observing two things—the outer brain activities via the mirror as well as the inner thought processes that no one else has access to. In order to make the connection between "inner" mental states and "outer" neural states, scientists would depend on information that only the "brain-owner" can provide. The world of the mind is accessible only to the "brain-owner," not brain scans. Even lie-detector tests don't detect thoughts, but at best the subject's physiological and emotional responses to his thoughts.

It seems rather obvious that there is no such thing as mind-reading through brain scans or other techniques. Contemporary neuroimaging techniques make it possible to observe directly only the effects of neurological activity such as changes in intracranial blood flow. But one cannot "see" cognitive activity itself, only the effects of cognitive activity. Consequently, neuroscientists cannot just "read" a person's mind. If they want to associate certain brain activities with certain mental activities, they need to ask their patients what they are thinking. So, again, the neural and the mental must be different from each other.

[177] Ludwig Wittgenstein, *The Blue and Brown Book* (New York: Harper and Row, 1980), 11–13.

Let us come to a conclusion. If one of the previous ten arguments convinces you completely, that would be enough. If they convince you only partially, then perhaps all of them combined provide enough compelling reason for you to reject the idea that the mental is identical to the neural, that the mind is identical to the brain, or that the soul is identical to the body.

Amazingly enough, many neuroscientists seem never to have heard of these arguments, ancient and modern alike, against their strongly held convictions. Or, having heard of them, they have either forgotten or willfully discarded them. They still think of the brain as an organ that secretes thoughts similar to the way the hypothalamus secretes hormones. Interestingly enough, as far back as the thirteenth century, St. Thomas Aquinas expressed his objection to this idea, saying, "For as intellect has no bodily organ, rational beings cannot be differentiated according to a physical diversity in the constitution of their bodily organs."[178] The brain may have gone through a process of evolution, but the mind did not. Some like to call the human mind the human soul; that's fine, as long as we realize that the mind is the intellectual part of the soul.

In this light, we can no longer say that the mind is identical to the brain. That would be like saying that the news report from a radio station comes from a receiver through its antenna, transistors, and speakers. These do indeed help transmit the news, but they are certainly not the origin of the news and do not determine the content of the news they broadcast. When the receiver breaks down, the news broadcast is still there. Therefore, it is better to say that the mind somehow "uses" the brain as a radio station uses the receiver. The brain acts as the physical carrier for the mind's immaterial thoughts. If something is broken in the carrier, that does not mean there is something broken in the mind.

[178] Aquinas, *De Anima*, 294

As William Shakespeare said in *King Lear*, "We are not ourselves when nature, being oppressed, commands the mind to suffer with the body." To be sure, thinking presupposes a functioning brain, but it cannot be reduced to this fist-sized organ. On the other hand, a malfunctioning brain would not directly affect the mind. A broken brain is as physical as a broken bone—but the mind is not physical, nor is the soul. Even people who have "lost their mind"—whether through dementia, Alzheimer's disease, or mental insanity—have not really lost their minds or their souls. The fact that they are human means that they have a human mind and a human soul. What they did lose somehow is a part of their brains, but not their minds. They have lost part of their "communication channel." A breakdown of the brain is not necessarily a breakdown of the mind.

The mind reflects God's Mind

First let's revisit Darwin's question: Can science be trusted if done by a human mind that science has found to be the result of natural selection? Again, the answer is no. We cannot trust science unless we are ready to derive the human mind (including its rationality and intellect) from a higher source. I would say the mind is the soul's eye, its light—or put differently, the mind is the intellectual part of the soul by which we know truth. Trust in science must ultimately be based on the power of the human mind.

Darwin could have cleared the confusion he had created for himself had he just acknowledged that the human mind is not a product of evolution and natural selection. The human brain (including its intelligence) may very well be a product of natural selection, but that doesn't mean the human mind (including its intellect) is too. As a matter of fact, the theory of natural selection must assume the human mind, but it can neither create nor explain it.

As we said earlier, to study the brain we need a mind, for no brain can study itself. So the mind must have another origin than the brain. I would even go further and claim that the mind must be something made in God's image, a reflection of the Creator's Mind. Whereas it was Darwin's conclusion that we cannot trust anything we know about God if our knowledge depends on natural selection, I would rather argue the opposite—that we cannot trust anything we know at all if there is no God.

When Charles Darwin wrote in an early private notebook, "Why is thought, being a secretion of brain, more wonderful than gravity as a property of matter?"[179] he was attacking the power of reasoning and logic by using the power of reasoning and logic—which sounds like a textbook case of circular reasoning. Instead, the power of the human mind can be understood only in reference to God.

If the human mind is more than the brain, then uniquely human faculties such as rationality and morality are not tied to the brain but to the mind. These faculties would not be determined by anatomy, physiology, genetics, and neuroscience, but instead they must come from the human mind—which is the intellectual part of the human soul. And in turn, mind and soul come directly from God—not from genes and neurons. So neuroscience cannot really be understood correctly without pointing to God.

The Bible told us this message already a long time ago, when God said, "Let us make man in our image, after our likeness."[180] So the human mind is a reflection of God's Mind. It wouldn't make sense to say that the human brain is a reflection of God's brain. Matter can evolve, but the human soul is immaterial and cannot evolve, but has to be created directly by God. The Catholic Church

[179] Paul H. Barrett et al., eds., *Charles Darwin's Notebooks, 1836–1844* (Cambridge, UK: Cambridge University Press, 1987), 291.

[180] Gen. 1:26a.

summarizes this as follows: "The Church teaches that every spiritual soul is created immediately by God — it is not 'produced' by the parents."[181] Put differently, it's not a product of evolution, genes, or neurons. The origin of the human soul and mind thus eludes science, which deals only with matter.

If anyone thinks that the mystery of the mind has been solved by science as a mere brain issue, it might be worth pondering the opposite — namely, that the secret of science is to be found in the mystery of the mind, the reflection of the Mind of the Creator. Only the mystery of the mind makes scientific knowledge possible—not the reverse. And the mystery of the mind can point only to God.

Seen in this light, humans end up being radically different from their "relatives" in the animal world. Although we are flesh as they are flesh, we are also very exceptional creatures in this universe, endowed with the faculties of language, rationality, morality, and religion. Genes do not determine which language we speak, which morals we have, or which religion we follow. In fact, they may not even determine that we do have the capacity to speak a language, possess a moral code, or follow a religion.

Human beings have got to be more than what science tells us; otherwise, science would in fact not have much to say. In addition to our five material senses, we have at least three immaterial "senses" — the rational sense of "true or false," the moral sense of "right or wrong," and the religious sense of "material or spiritual." The origin of these three immaterial senses seems to keep eluding us when we leave it up to science. However, we should not forget that neuroscience is impossible without the minds of neuroscientists. Ironically, neuroscience tends to deny what it badly needs for its own existence — the human mind. Whether neuroscientists like to admit it or not, even neuroscience points ultimately to God.

[181] CCC 366.

Although there may be some random meandering in the process of evolution, as we found out earlier, its end goal is set. The evolutionary process is aiming for the highest level of existence in the universe: the rise of humanity; the rise of persons who are rational, moral, and religious; the rise of beings made after the image and likeness of God. That outcome would be sheer magic without God.

9

How Behavioral Science Points to God

Unlike most plants, animals show some kind of social behavior, as do humans—that is, they interact extensively with each other. But there is (at least) one thing that sets human behavior apart from animal behavior: humans show *moral* behavior. Evidently, there are two distinct kinds of human behavior—social behavior and moral behavior. The difference is basically that social behavior is about how humans *do* behave, while moral behavior is about how they *should* behave.

To put it another way: on the one hand, there is a regular kind of human behavior, which we find also in the animal world: we do things to others and others do things to us. On the other hand, there is also a unique kind of human behavior, very different from regular, social behavior—it's called moral behavior: we *ought* to do certain things to others, and others *ought* to do certain things to us. We *owe* certain actions to others, and others *owe* them to us. In more technical terms: we have *rights*, which others owe us, and we have *duties*, which we owe to others.

Behavioral scientists tend to focus exclusively on regular human behavior—what humans do to each other—but many of them have a blind spot for moral behavior. However, leaving moral behavior out of the picture amounts to a degradation of human behavior. We end up with only part of the whole story. The other part is often conveniently concealed in another field called ethics.

What is so peculiar about moral behavior?

Whereas social behavior describes merely what we do to each other, *moral* behavior dictates what we *should* do to each other. That makes human behavior unique, unknown in the rest of the animal world. Access to a moral code, to a world of "oughts" and "rights" and "duties," seems to be a unique faculty of human beings, separating them from the rest of the animal world. So far in this book, we have been judging things in terms of *rationality* as "true or false," but now we are judging things also in terms of *morality* as "right or wrong." How do we know what's right and what's wrong?

Ordinary people with common sense know right from wrong; that's one of the reasons why they can qualify as jurors in the judicial system. Even though some human beings do kill each other, murderers still know, deep down, that what they are doing is wrong, even as some passion such as greed, revenge, anger, or hate, overtakes their moral common sense. This "common sense" is something all humans possess because it comes with human nature, with being human, regardless of ethnicity and race. There seems to be some kind of general moral law that we all share.

How are we to take that general moral law? Think of this: when someone denies the existence of the laws of nature, such as the law of gravity, all one can do is invoke the principle of common sense. Common sense tells us there is some kind of *physical* order in nature. Stones that fall today will also fall tomorrow. We cannot prove this today, but we can confidently assume it for tomorrow. Similarly, when someone denies the existence of moral laws, all one can do is invoke the principle of common sense again. Common sense tells us there is some kind of *moral* order in life: if murder is wrong today, then it will also be wrong tomorrow.

The question is, though, where moral laws come from. Where do they reside in reality? Let us find out first what does *not* qualify as a basis of morality before we try uncovering its real basis. Unfortunately, there have been many trials to reduce morality to something amoral—perhaps as something real but certainly not moral.

Here is trap number one: morality comes from the animal world. The fact is that animals do not and actually cannot have morality. Animals do engage in social behavior, but they don't have moral behavior regulated by a moral code. The relationship between predator and prey, for instance, has nothing to do with morality; if predators really had a conscience guided by morality, their lives would be pretty harsh. As a consequence, animals never do awful things out of meanness or cruelty, for the simple reason that they have no morality and thus no cruelty or meanness. They follow whatever pops up in their brains—and no one has the right to blame them morally. Dogs may act as if they are caring, but they are just following their instincts, not some moral code. Dogs happen to have an instinct to act friendly, whereas cats lack it, since it is not in their genes. Dogs are known for their social behavior, but don't confuse this with moral behavior.

Then there is trap number two: morality comes from our genes. How could that be? DNA is physical "stuff" that can be long or short, light or heavy, but morals cannot be any of these—they have no mass, size, or color. Besides, if morality were in our genes, why would we need an articulated moral code to reinforce what "by nature" we would or would not desire to do anyway? Under such circumstances, a moral code would be completely redundant. Instead, the opposite could be argued: morality has the power to overrule what our genes "dictate"—passions, emotions, and drives.

As a matter of fact, far too many people are willing to break a moral rule when they can get away with it. It is hard to believe that they are going against their genes. A genetic code would make us act a certain way "by nature," whereas a moral code would make us do "by choice" what our genes do not make us do "by nature." Bad moral behavior can spread like wildfire, but mutated genes don't spread that quickly. Instead we have a moral code because God, according to St. Augustine, "wrote on the tables of law what men did not read in their hearts."[182]

Then there is trap number three: morality is the product of natural selection. That's hard to believe when you think about it. Moral laws such as "You shall not kill" or "You shall not commit adultery" could hardly make it through natural selection, for their offenders—the killers, the promiscuous, and the rapists—would do much better in reproduction than their victims or than those who follow a moral code. Moral laws do not seem to have much survival value and, therefore, are not good candidates for natural selection.

Whereas natural selection is based on self-preservation at the cost of others, morality is often self-sacrifice for the good of others. When firefighters or soldiers on the battlefield die in the line of duty, they typically do not give their lives for the sake of natural selection. Francis Collins, the former head of the Human Genome Project and currently director of the National Institutes of Health, made very clear that morality goes against natural selection: "Evolution would tell me exactly the opposite: preserve your DNA. Who cares about the guy who's drowning?"[183]

Then there is trap number four: morality comes from past experiences. What is wrong with that claim? Well, killing is morally

[182] Augustine, *Enarrationes in Psalmos* 57, 1. Quoted in CCC 1962.

[183] Francis Collins, "An Interview with Francis Collins," *The Question of God*, PBS, 2004.

wrong—but certainly not because we discovered so after we had killed some people or had seen some killings. That would mean we would have to do something wrong or watch others do something wrong before we could know what is wrong. A moral command comes before what it commands, not after. Morality may be corroborated by past experiences, but it is not created by such experiences—in fact, it aims to prevent them.

Besides, how would we know that what we did or saw was wrong? Knowing what is right and wrong must come *before* we do or see something wrong. Although we may suppress it, we must have some moral knowledge to begin with. There is mounting evidence that babies as young as six months old make moral judgments. Researchers have found that, even if an experiment is unfairly rigged so that one child receives more rewards, children will ensure that a reward is fairly split, whereas animals usually fight for the largest piece.[184]

Then there is trap number five: morality is something acquired—through upbringing, training, discipline, or education. No doubt, discipline and education are part of dealing with morality. People who are at the mercy of their lusts, drives, and passions may not do the good they ought to do, because they are not disciplined enough to resist their lusts. But that does not mean that morality is merely a matter of being educated, taught, and disciplined. The laws of nature, such as the law of gravity, may have to be taught to us in a physics class, but they are not a product of training and teaching. It is partly through schooling that we know about them, but the laws themselves are not a product of schooling—they existed already before we learned about them.

Something similar can also be said about moral laws. While parents may help us understand moral laws and may have prepared

[184] J. K. Hamlin et al., "How infants and toddlers react to antisocial others," *PNAS* 108 (2011): 19931–19936.

us to do what is morally right, the distinction between right and wrong is not a matter of upbringing. Slavery, for example, was wrong long before we knew or recognized it as wrong. In other words, a moral code is not the product of upbringing, but its standard. If morality were a product of upbringing, that would still raise the question as to where it began and how. That's basically an endless pursuit.

Then there is trap number six, a very common trap: morality is a matter of intuition. First of all, this opens the argument up to the attack that morality is not real, but only exists in a person's mind—a thought famously expressed as "many heads, many minds." George Bernard Shaw, for instance, spoke of "different tastes."[185] If morality were merely a matter of intuition or taste, no further discussion would be possible. The best we could say is that some people have better taste than others.

However, what is morally good is not a matter of what *feels* good. Feelings can never be the standard for judging morality, for we would have to decide next who has the best "gut feelings." It is actually the other way around: a moral code is the standard for judging feelings. Morality determines which moral intuitions are right or wrong. Otherwise, all defendants in court would be entitled to claim that they only followed their "gut feelings."

Then there is trap number seven: morality is a matter of conscience. Ironically, even moral relativists, who deny that morality has any absolute authority, still hold on to at least one moral absolute: "Never disobey your own conscience." Those who say so should at least ask themselves where the absolute authority of a human conscience comes from. We cannot validly justify that our act was morally right by claiming that our conscience tells us

[185] George Bernard Shaw, *Man and Superman* (New York: Penguin Classics, 2001), 211.

so. The idea that one's conscience creates moral laws is as flawed as the idea that one's consciousness creates the laws of nature.

If conscience determines what is good or bad, then we must call the Nazis "good" people because they just followed their conscience, or we must call slaveholders "good" people because they only followed their conscience. Or take two opposing armies in a war: they both use their conscience to claim that they are right. That position is hard to defend, for the claims contradict each other. Both cannot be right at the same time. There must be something above and beyond conscience that determines what is morally right or wrong.

So where does this leave us? If we are looking for a key to understanding moral behavior, this key will not be found in something material, such as genes, but in something immaterial: the mind. If morality comes neither from the animal world, nor from genes, nor from experience, nor from upbringing, nor from intuition, nor from conscience—where else could it come from? Could it be that moral behavior can only be understood as pointing to God?

How do we know what's right and wrong?

Who or what can tell us what we *ought* to do? Why should whites care about blacks, or the rich about the poor, or the strong about the weak? There is nothing in another human being that forces me to be "good" to him. No society or government has the right to demand my absolute obedience in dealing with others. No other human being has the right to demand absolute obedience from me. No one, not even I myself, has the right to demand my absolute obedience.

The only authority that can require me to do what I should do is someone infinitely superior to me; no one else has that much authority. In other words, there is only one reason for us to do good and be good to others—the reason being that God is our Father who created us all as His children, made in his image and likeness. Only God can demand absolute obedience from us.

How do we know, then, what God demands of us morally? Let's call in common sense again. Common sense tells us that there are things all of us know we ought or ought not to do, for the simple reason that we are human beings endowed with the faculty of morality, in addition to the faculty of rationality. There are moral laws that should not be ignored, just as there are laws of nature that cannot be ignored. There is a moral order in nature as much as there is a physical order. The main difference is that physical laws *can't* be ignored, whereas moral laws *shouldn't* be ignored.

Somehow, we "know" what is morally right or wrong. How can every human being know all this? Again, the basic answer is common sense, something that comes with human nature and is shared by all of humanity. True, there are some important disagreements about what exactly is "good" among different cultures, but beneath all disagreements about lesser moral laws and values, there always lies an agreement about more basic ones. Peter Kreeft compares this with different languages: beneath the different words of different languages you find common concepts—this is what makes translation from one language to another possible. In the same way, he concludes, "we find similar morals beneath different mores."[186]

In fact, there is not a great deal of difference between Christian morality, Jewish morality, Hindu morality, Muslim morality and Buddhist morality, although there is a great difference in the religions themselves. C. S. Lewis was even able to publish a list of

[186] Peter Kreeft, *A Refutation of Moral Relativism: Interviews With an Absolutist* (San Francisco: Ignatius Press, 1999), 70.

universal moral principles, which he called "Illustrations of the Tao or Natural Law."[187]

We find this idea of commonality also in St. Paul's reference to pagans "who never heard of the Law but are led by reason to do what the Law commands."[188] We find it in the work of the first Christian philosopher, St. Justin the Martyr, who wrote around the year 150 while living in the turbulent, mostly pagan Roman Empire.

> Every race knows that such things as adultery, and fornication, and homicide are sinful. For example, though they all fornicate, they do not escape from the knowledge that they are acting unrighteously—with the exception of those possessed by an unclean spirit, those debased by wicked customs and sinful institutions, and those who have quenched their natural ideas. For we observe that such persons refuse to endure the same things they inflict on others. They also reproach each other for the evil acts that they commit.[189]

It is mostly thanks to St. Thomas Aquinas that this concept of moral commonality has become known as the "natural law." Its key idea is that moral laws are based on human nature, on the way we *are*. As a consequence, morality is a function of human nature, so that reason can discover valid moral principles by looking at the nature of humanity and society.

This means that what we ought to *do* is related to what we *are*. "Thou shalt not kill," for instance, is based on the real and objective value of human life and the need to preserve it. "Thou shalt not commit adultery" is based on the real value of marriage, family, and mutual self-giving love, and on children's need for trust and

[187] In the appendix of his book *The Abolition of Man* (1944).
[188] Rom. 2:14, Jerusalem Bible.
[189] Justin, *Dialogue with Trypho*, chap. 93.

stability. We share these moral convictions, to a large degree, with all of humanity. Every culture in history has had some version of the Ten Commandments—they are part of the moral code humans ought to follow by virtue of being humans with moral behavior.

Only the commonality of the natural law can explain that all humans have certain moral rights and duties. Rights are what other human beings morally owe us and *ought* to render to us; duties are what we morally owe other human beings and *ought* to render to them. Moral rights and duties are *universal*—that is, the same for everyone everywhere, regardless of race, ethnicity, nationality, culture, or political affiliation. Consequently, the natural law is common property of all human beings and is by its very nature not connected with interest groups or majority consensus or particular religions. The *Catechism* confirms this: "The natural law, present in the heart of each man and established by reason, is universal in its precepts and its authority extends to all men."[190] Although the natural law may not be universally obeyed, or even universally admitted, it is still universally binding and authoritative.

Not surprisingly, St. Thomas Aquinas made a clear distinction between the universal natural law and the laws (legal, civil, or positive) made and upheld by local governments. That is the reason why Martin Luther King Jr. could say, "A just law is a man-made code that squares with the moral law, or the law of God. An unjust law is a code that is out of harmony with the moral law. To put it in the terms of Thomas Aquinas, an unjust law is a human law that is not rooted in eternal and natural law."[191]

Interestingly enough, without the universality of the natural law, there would not have been any justification for the Nuremberg

[190] CCC 1956.

[191] Martin Luther King Jr., "Letter from Birmingham Jail," April 16, 1963.

trials that took place after World War II—or for any other international court, for that matter. From a purely legal point of view, it would not have been right, or even possible, to bring to trial and punish the Nazis who had applied the civil laws that were created and implemented by a regime that had come to power through legal channels—making them "law-abiding" citizens following the law of the land. But from the perspective of natural law, their "lawful" actions were in fact atrocities committed against humanity.

As we said earlier, the only authority that can require me to do what I ought to do is someone infinitely superior to me; no one else has the right to demand my absolute obedience. This means that there are no rights and no duties without God. An absolute law can come from and be enforced by an absolute Will alone. The writer Fyodor Dostoyevsky had it right when he said that without God, all things are permitted.[192]

Even the atheist French philosopher Jean-Paul Sartre realized that there can be no absolute and objective standards of right and wrong if there is no eternal Heaven that would make moral laws and values objective and universal. He had to conclude that it is "extremely disturbing that God does not exist, for there disappears with him all possibility of finding values in an intelligible heaven. There can no longer be any good *a priori*, since there is no infinite and perfect consciousness to think it."[193] Because Sartre denied the existence of God (until just before his death), he realized very clearly that his atheism required him to give up on morality. If there is neither a God nor any eternal goodness, there cannot be evil either. As Thomas Aquinas famously said,

[192] Fyodor Dostoyevsky, *The Brothers Karamazov*, bk. 11.

[193] Jean-Paul Sartre, *Existentialism Is a Humanism* (New Haven, CT: Yale University Press, 2007), 28.

"Good can exist without evil, whereas evil cannot exist without good."[194]

The German philosopher Friedrich Nietzsche was another atheist who realized how devastating the decline of religion is to the morality of society. He wrote, "God is dead; but as the human race is constituted, there will perhaps be caves for millenniums yet, in which people will show his shadow."[195] Nietzsche is saying here that humanism and other "moral" ideologies shelter themselves in caves and venerate shadows of the God they once believed in; they are holding on to something they cannot provide themselves. They have constructed "idols" to preserve the essence of morality without its substance.

Nietzsche clearly understood that "the death of God," as he called it,[196] means the destruction of all meaning and value in life. He saw that neither our dignity nor our morality would be able to survive in a world without divine and eternal laws. Jürgen Habermas, a nonreligious philosopher, expressed his conviction that the ideas of freedom and social coexistence are based on the Jewish notion of justice and the Christian ethics of love: "Up to this very day there is no alternative to it."[197] This does not mean, of course, that we must believe in God in order to live a moral life. As Nietzsche put it, we can still venerate "shadows from the past."

Because of all of this, we cannot but recognize that morality comes from "above." Moral laws, moral values, moral rights, and moral duties ultimately reside in Heaven. They are real because they come from the One who created this world. God gave the world not only

[194] Aquinas, *Summa Theologica* I, q. 109, art. 1.

[195] Friedrich Nietzsche, *The Gay Science*, trans. Walter Kaufmann (New York: Vintage Books, 1974), § 108.

[196] Ibid., § 125.

[197] Jürgen Habermas et al., *Time of Transitions* (Cambridge, UK: Polity Press, 2006), 150–151.

laws of nature but also laws of right and wrong. Therefore, we ought to do what we ought to do—for Heaven's sake! The United States' Declaration of Independence got it right: we are endowed by our Creator with certain unalienable rights. When, in 1948, the United Nations (UN) affirmed in its Universal Declaration of Human Rights that "all human beings are born free and equal in dignity and rights," it must have assumed the same—that such rights are not man-made, but God-given—though the drafters famously left the term "right" vague in order to achieve passage. The Catholic philosopher Jacques Maritain, who was actively involved in drafting the U.N. declaration, said, paradoxically, "We agree on these rights, on condition that no one asks us why."[198] If we do ask why we have these human rights, there is only one reason: God has endowed us with them.

Therefore, without a firm foundation in God, "equality in dignity and rights" would be sitting on quicksand, subject to the mercy of law makers and majority votes. It is through the "voice" of God in the natural law that we know about right and wrong, about human rights and human duties. Without God, who is the Author of human rights, we would have no right to claim any rights. If there were no God, we could not defend any of our rights. We would have at best (legal) *entitlements*, which the government provides, but no (moral) *rights*, for only God can provide those.

Entitlements are at best privileges others give us, not rights that God alone can give. John F. Kennedy said it well in his inaugural address: "The rights of man come not from the generosity of the state, but from the hand of God." Without God, there could be no absolute or objective standards of right and wrong. If these did not come from God, people could take them away at any time—which they have certainly attempted numerous times.

[198] Jacques Maritain, *Man and the State* (Chicago: University of Chicago Press, 1951), 90–100.

Apparently, we have an important distinction here: *rights* are God-given, while *entitlements* are man-made. Some people think of human rights as if they were entitlements that the government gives us. Indeed, we gain entitlements as we age—in the US, one can drive a car at sixteen, vote at eighteen, and buy alcohol at twenty-one. But we cannot apply this kind of reasoning to human rights. We have rights because all human beings are God's creatures subjected to His natural law, whereas entitlements we have only because we belong to a certain society. Rights are God-given, such that we cannot invent them on our own, while entitlements are something that individual societies invent and promulgate all the time. The government can hand out entitlements, but it cannot give rights away, although it may sometimes try to take them away.

Unlike entitlements, moral rights and duties are universal, absolute, timeless, objective, and nonnegotiable standards of moral behavior. The *Catechism* puts it this way: "The natural law is a participation in God's wisdom and goodness by man formed in the image of his Creator. It expresses the dignity of the human person and forms the basis of his fundamental rights and duties."[199] Therefore, when it comes to morality, we cannot just pick whatever we want. Morality obliges us to pursue unconditionally what is good and right.

Because there is no morality without God, moral behavior points to God as the moral Lawgiver. Morality can find its foundation only in God, so any morality without God would be floating in midair. This doesn't mean, of course, that people who do not believe in God cannot act morally. They surely can, for most people do condemn murder, whether they have religion or not. Nietzsche would probably say that they shelter themselves in caves and venerate shadows of the God they once believed in.

[199] CCC 1978.

Obviously, one doesn't have to be a Christian to do good and be a good person. Of course, you can be moral without knowing why you are a moral being and why you ought to be moral. But only belief in God can explain to us why we *ought* to do good, why we have the moral *duty* to do good. The answer is that we owe it to God and, therefore, to each other. Moral behavior has its roots in Heaven. It is actually almost unthinkable for any behavioral science to ignore this moral part of human behavior and deny that it points to a Lawgiver, God.

What does conscience have to do with this?

When it comes to moral behavior, human behavior is strongly associated with a moral *conscience*. For some enigmatic reason, the authority of a person's conscience still ranks high in people's minds. Conscience is now the highest court of appeal—it has been given ultimate "primacy," coming close to infallibility. It turns out that almost all people have something about conscience that they respect, even if their theory is that conscience is nothing. What, then, has made the term "conscience" so popular? Probably the main reason is that the slogan "Follow your conscience" has come to be code for pursuing one's personal preferences and desires. And who would not like that?

Earlier, we argued that morality cannot be based on conscience only. So we need to explain what is wrong with the slogan "Always follow your conscience." Well, there is a lot of confusion and ambiguity behind the idea of "just following your conscience." We need to find out first what a conscience is.

Our conscience has often been compared to technical devices we are all familiar with: a compass, a global positioning system (GPS), a barometer, an alarm, a gas gauge in a car—the list goes on and on. What these analogies get right is that our conscience is indeed a monitoring device—it monitors what is good or bad, right or wrong. What they mask is the fact that the aforementioned devices are merely tools that may not work properly and may even fail entirely. And they certainly cannot work on their own.

A real compass, for instance, functions as a pointer to the magnetic north, because the magnetized needle aligns itself with the lines of the earth's magnetic field—that is, with something outside itself. But it should not be used in proximity to ferrous metal objects or electromagnetic fields, as these can affect its accuracy. At sea, for example, a ship's compass must be corrected for errors, called deviations, caused by iron and steel in the ship's structure and equipment. Besides, the compass itself may have a defect. The needle on the gas gauge in your car, for instance, may no longer go down because it is broken, yet the tank may be almost empty. Your GPS may not work when something obstructs the connection with the satellite high above your head.

In other words, a person's conscience may indeed function like a compass or GPS, but these "monitoring" tools must themselves be monitored and aligned with an outside source. The same can be said of our conscience. Just as a compass needs to be aligned with the earth's magnetic field and protected from surrounding interference, and a GPS needs to be tuned to the right feed from satellites high in the sky, so does a human conscience need constant alignment.

But keep this in mind: the fact that our moral compass may sometimes fail does not mean that there is no right direction at all. As in math, we may get our sums wrong, but it does not follow that a correct answer does not exist. The question, then, is this: what is the "right math" in morality? What is the right feed for our conscience? How do we properly align it, and to what? In short, how do we calibrate our conscience?

The Catholic Church would say that human beings are created with a moral compass pointing not to the magnetic north, but to the "Above"—to a place where justice reigns and moral laws reside. Therefore, our conscience is not a private "compass" that determines its own north pole; it has to be aligned with the one real "North Pole Above"—otherwise, we can easily go off track.

If our moral compass is off track, our conscience can basically tell us anything. The political philosopher Hannah Arendt observed that "just as the law in civilized countries assumes that the voice of conscience tells everybody 'Thou shalt not kill' . . . so the law of Hitler's land demanded that the voice of conscience tell everybody: 'Thou shalt kill.'"[200] Such things happen when our moral compass becomes a private compass that determines its own direction.

So we must come to the conclusion that there is more to moral behavior than having a conscience and following it. When people say, "Never disobey your own conscience," they forget that one can do things "in good conscience," but also "with a bad conscience." Therefore, a conscience on its own can be good as well as bad. Many people are unaware that they may have a moral compass that is broken, or they may not even realize that they have a moral compass at all. They end up following their genitals in sexual affairs, their curiosity in biomedical research, or their personal desires in matters of life and death—no further questions asked. That amounts to having human behavior without moral behavior. However, personal desires cannot possibly be the source of morality. It should be the other way around: morality should judge our desires.

As a consequence, someone's conscience cannot have absolute authority in and of itself. Our conscience does not create moral laws and values, but merely receives them. A person's conscience does not speak on its own, but reflects the natural law bestowed on us by God. This is why we cannot take our conscience as an

[200] Hannah Arendt, *Eichmann in Jerusalem: A Report on the Banality of Evil* (New York: Penguin Classics, 2006), 150.

entirely private issue that we can form at our discretion. To use another analogy, a compass does not create its own magnetic field. Therefore, a person's conscience is not the highest moral authority there is, as it is subject to the supreme authority of the natural law, which comes directly from God.

As Vatican II puts it, "In the depths of his conscience, man detects a law which he does not impose upon himself, but which holds him to obedience."[201] The *Catechism* calls our conscience "man's most secret core and his sanctuary. There he is alone with God whose voice echoes in his depths."[202] Therefore, when people follow their conscience, it is important that they listen to God's voice, not their own. In the words of emeritus Amherst College professor Hadley Arkes, "Conscience is not directed inward to oneself and one's feelings, but outward to the natural law and moral truths."[203]

The fact that we can be morally blind—blinded by upbringing, culture, character, personality, temperament, lust, or lack of discipline—explains why we need help to correct a faulty conscience. Our conscience should be in a perpetual dialogue with God. A "dialogue" with oneself amounts to a monologue that isolates and alienates us from God, our moral Lawgiver. The one who is supposed to follow the law has no right to make himself the lawgiver by becoming a sovereign individual in the place of God and His Church—this would be a form of idolatry. The only authority that can obligate us is someone infinitely superior to us; no one else has the right to demand our absolute obedience. Archbishop Fisher puts it this way: "A Catholic must be prepared to accept moral

[201] Vatican Council II, Pastoral Constitution *Gaudium et Spes* (December 7, 1965), no. 1.

[202] CCC 1776.

[203] Hadley Arkes, "Recasting Religious Freedom," *First Things* (June 2014): 47.

instruction from the Church and never appeal to conscience to make an exception for himself."[204]

Our conscience is like an alarm that alerts us before we sin; when it goes off, we must not ignore it. When a red warning light in our car lights up, we should have the problem fixed—not by disconnecting the light, but by fixing what causes it to light up. The same goes for our conscience: we must not silence it. However, when the alarm does *not* go off, this does not mean that there is an "all-clear" sign, for we may have intentionally ignored its upkeep. That is how we can willfully manipulate or even damage our conscience. It requires "maintenance service" and needs to be "calibrated" again and again, often with help from the Church.

So we must come to the conclusion that behavioral science cannot neglect moral behavior, and that moral behavior in particular points to God for its validation and foundation. We cannot explain moral behavior in terms of science. Instead, science itself must be judged by a moral code, for not everything that is scientifically possible is also morally permissible. Albert Einstein was right to speak of "the moral foundations of science." He added, "You cannot turn around and speak of the scientific foundations of morality.... Every attempt to reduce ethics to scientific formulae must fail."[205]

This means that all human behavior is subordinate to the moral code of moral behavior. A moral code cannot be fully understood without reference to God, for without God, everything is permitted, even in science. Morality can interrogate science, but science cannot question morality—it is beyond its reach.

[204] Anthony Fisher, "Struggling to Recover a Catholic Sense," Zenit, March 3, 2007, § 2.3.

[205] In a discussion on science and religion in Berlin in 1930.

Morality cannot be properly understood without pointing to God. Our task isn't to decide what's right and what's wrong—God settled that at Creation. Our task is to do what's right and avoid what's wrong.

10

How Semantics Points to God

All languages use words to make sentences. Most words have a meaning—they "mean" something to us and to other users of the same language. Semantics, a field of linguistics, is the science of meaning. It studies the relationship between words and what they refer to, or their *reference*. What a word refers to depends on its meaning, which is determined by the *concept* behind that word.

Concepts usually have a web of connections with other concepts. The concept of "circle," for example, has an intricate web of connections with other concepts such as "radius" and "diameter." Or take the word "blood": its underlying concept—its meaning, if you will—is connected with other concepts such as "hemoglobin," "red blood cells," "bone marrow," and the like. Concepts determine what a word refers to. The concept of "blood," for example, makes the word "blood" applicable to humans, apes, and all other mammals.

Concepts are the building blocks of sentences. For instance, the sentence "The enzyme amylase breaks starch down into sugar" contains concepts such as "enzyme," "amylase," "starch," and "sugar." We can understand this sentence only if we understand the concepts that are in it. So concepts are essential for our understanding.

Concepts in the animal world?

Evolutionary biology has led some, especially in the neo-Darwinian camp, to believe that animals are "humans in the making" and that humans are "glorified animals" with only small transitions separating them.[206] It's a conviction that makes these believers claim that concepts are not exclusive to humans, but must exist in the animal world too. There are several reasons, though, why it's very questionable whether animals are capable of using concepts.

First of all, concepts are more than categorization tools. Animals are certainly capable of classifying and categorizing—they distinguish predators from prey, females from males, animals from plants, and so on. But does that mean animals have concepts—the concept of a "predator" or a "female" or a "plant," for instance? Not necessarily. Many concepts do indeed classify and categorize, but we have to keep in mind that classification and conceptualization are not identical. Obviously, in order to conceptualize, we often need to categorize, but the reverse is not true: in order to categorize, we do not need to conceptualize. Conceptualization goes much further than classification. Concepts are powerful tools only humans use to see similarities even when they are not directly visible—take, for instance, the similarities between herbivores, a classification beyond what is directly visible.

Let's take buffalos as another example. They do not identify a lion with the concept of "predator," but as something to run away from—and they will do the same with other predators. That's an inborn or learned response to an acute situation. The similarities animals see are not connected with concepts but with similar stimulus-response reactions. They learned, or were probably genetically programmed, to make certain associations—no concepts needed.

[206] Technically, humans are a special kind of animals (*animal rationale*), but I will use the term "animal" only for non-human animals.

Second, concepts are different from associations. Animals do have the ability to learn to associate a certain input, a stimulus, with a certain output, a response. Stimulus-response associations are very common in the animal world. Such associations have been extensively used to train animals and to perform behavioral experiments. Humans have them, too, but associations are not related to concepts. Animals don't have mind-dependent concepts but only a series of mind-independent associations between objects in the external world and the signals taught to them. These signals have a reference, but no meaning, whereas concepts have meaning, but not always a reference. Even chimpanzees—however "advanced" as they seem to be in terms of being humans in the making—are perfect examples of what the scientists Berwick and Chomsky call pure "associationist learners," who make direct connections between particular external stimuli and their signals.[207]

Even Jane Goodall, the closest observer of chimpanzees in the wild, had to come to the conclusion that for chimpanzees, "the production of a sound in the *absence* of the appropriate emotional state seems to be an almost impossible task."[208] What makes associations so different from concepts is the fact that concepts have no necessary link to a particular situation. We can talk about a "dinosaur" without ever having met one in real life. We can talk about a "proton" without ever having seen one. We can talk about "tomorrow" without having access to the next day ahead of time.

Third, concepts are different from labels. True, animals may very well be able to use labels for certain objects. Sarah the chimp,

[207] Robert Berwick and Noam Chomsky, *Why Only Us: Language and Evolution* (Cambridge, MA: MIT Press, 2016), 146.

[208] Jane Goodall, *The Chimpanzees of Gombe: Patterns of Behavior* (Boston: Belknap Press, 1986), 125.

for example, was trained to use cards for communication.[209] But whether she also used concepts is very doubtful. No ape has ever been seen using such labels other than as links associated to concrete objects in their surroundings. Herbert Terrace and his team made the chimp they had named "Nim Chimpsky" learn 125 signs in American Sign Language,[210] but while they taught Nim to use the label "apple," for instance, it's very questionable whether he had the concept of "apple."[211]

As Laura Petitto observes, "Chimps, unlike humans, use such labels in a way that seems to rely heavily on some global notion of association. A chimp will use the same label *apple* to refer to the action of eating apples, the location where apples are kept, events and locations of objects other than apples that happened to be stored with an apple (the knife used to cut it)."[212] In other words, the label "apple" was used to refer to *anything* associated with apples. But that doesn't make the label a concept.

Fourth, concepts are different from perceptions and sensations. Most sensations come in through the eyes, especially for vision-oriented animals. But images are not concepts. No matter how many cows, horses, and goats one has seen on the retina, the concept of "herbivore" does not spontaneously emerge from these sensations. Undoubtedly, animals do have the ability to see similarities by mere perception; that's how they can identify,

[209] David Premack and Ann James Premack, "Teaching Language to an Ape," *Scientific American* 227 (October 1972): 92–99.

[210] Herbert Terrace et al., "Can an ape create a sentence?," *Science* 17, no. 4 (1979): 396–410.

[211] For a good critical analysis, see Herbert Terrace, *Nim: A Chimpanzee Who Learned Sign Language* (New York: Alfred Knopf, 1979).

[212] Laura-Ann Petitto, "How the brain begets language," in *The Cambridge Companion to Chomsky*, ed. James McGilvray (Cambridge, UK: Cambridge University Press, 2005), 84–101.

recognize, and categorize food, predators, and mating partners. They may even be able to recognize a circle after being repeatedly presented with a circular object, which could be considered a primitive form of abstraction.

However, their seeming capacity for abstraction is closely tied to what the philosopher Mortimer Adler calls a "perceptual act."[213] In experiments, some species of fish, for instance, were able to recognize a circle, but only when presented with an actual circular object.[214] So this kind of abstraction could rightly be called "*perceptual* abstraction" because it is closely tied to actual perception, and does not work apart from that perception.

Humans, on the other hand, can also see similarities through concepts—that is, apart from any perceived object. Only human beings can think about a circle or about circularity in general, apart from any specific perceived object. That's what we do in geometry, for example. We could call this, with Adler, "*conceptual* abstraction" to distinguish it from "perceptual abstraction." Adler maintains that that there is no scientific evidence that any animal other than human beings can understand universals, such as a circle, detached from particular sensations.[215]

Fifth, concepts are different from signals. Words can be used either as signals—referring to physical entities—or as symbols—referring to mental entities such as concepts. Take, for instance, the word "poison": it can be either a symbol referring to a concept that explains its dangerous nature, or it can be a signal or a label on a bottle that alerts us not to take that stuff. Signals are very different from symbols.

[213] Adler, *Intellect*, chap. 4.

[214] Barr, *Modern Physics*, 192.

[215] Mortimer J. Adler, *The Difference of Man and the Difference It Makes* (New York: Fordham University Press, 1993), 136–138.

This difference, again, separates the world of animals from the world of humans. Humans can use signals and symbols alike, but animals can use signals only. Animals treat everything in their surroundings as signals that call for a direct response—by way of association—but they cannot use concepts to ponder realities beyond their needs for food and sex. They act mainly by instinct, not by using concepts. When birds are having an exchange of vocal signals in the morning, they are not having an exchange of ideas. Signals refer directly to a specific thing or situation, whereas concepts usually do not.

When prey animals flee from a predator, they are not reading the predator's mind, but processing certain signals. A prey animal, for instance, can take a "predator" only as a signal to flee or attack, but not as a symbol that can carry various interpretations. Only humans can think about a predator as an animal in need of food, as an animal who is born to prey on others, as an animal trained that way, as a pre-programmed killer, as an inevitable part of life, or as a part of nature that needs to be preserved. Humans can come up with such different interpretations, but animals cannot, for animals see only signals that call for an immediate response.

Sixth, concepts are different from warning signs. Vervet monkeys, for instance, have been shown to use different warning signs for different types of enemies.[216] But these warning signs are very different from concepts. These monkeys use their different warning signs only when a particular enemy is around; they do not use them simply to "ponder" and "chat" about their enemies in general during their spare time. Signals are situation-specific. Again, animals treat everything in their surroundings as signs that call for an immediate, direct, and definite response.

[216] Robert M. Seyfarth et al., "Vervet monkey alarm calls: Semantic communication in a free-ranging primate," *Animal Behaviour* 28, no. 4 (1980): 1070–1109.

When we train dogs to associate a command such as "The boss!" with their real boss, then the dog has been conditioned to respond to such a command by looking for the real boss. The command has become a sign. The dog has a physical image of its own boss, but it has no concept of what a "boss"—any boss, for that matter—is like. Humans, on the other hand, can use the word "boss" also as a symbol—a universal concept of "any boss" in general. They often use that word to talk about what their own boss is like, or should be like—preferably only when their boss is not actually present. Whereas signals call for direct action, concepts do not.

Seventh, concepts are more than communication tools. Scientists like to mention the famous example of a bee dance, which communicates to other bees where to find nectar.[217] Every movement by the dancer has a "meaning" for the other bees. They can tell the distance of the food source by her wiggling abdomen: the greater the distance, the more slowly she wiggles. The direction of the food is revealed by the direction and angle the dancing bee cuts across the circle. If she wiggles across the circle straight up, the watching bees know they will find the food by flying toward the sun; if she cuts the circle straight down, they know they have to fly away from the sun; if the dancing bee cuts across the circle at an angle, the other bees know they must fly to the right or left of the sun at the same angle. It's a pretty fancy communication system, but again, it doesn't work with concepts.

Every movement of the dance has, indeed, a certain "meaning" in terms of communication. But it's not the meaning that comes with concepts. The meaning of the bee dance is very concrete, concerning specific food at a specific location. There is nothing universal or abstract about it, as is the case with concepts. In other

[217] Karl von Frisch, *The Dance Language and Orientation of Bees* (Cambridge, MA: Belknap Press, 1967).

words, bees don't have a concept of food, not even nectar; they don't have a concept of geometry or circles; they have no concepts at all, period. They are programmed for how to communicate, but have no "idea" what they are communicating because they have no concepts.

Apparently, we just keep finding more and more reasons why there is no indication that concepts exist in the animal world. After all we have seen so far about the difference between concepts and what animals are able to do with what only appear to be concepts, we must come to the conclusion that there is a deep divide between the human world of concepts and the animal world of associations, labels, signals, and signs.

What concepts are not

Since there seems to be no way of tracing concepts back to the animal world, we need to look in the human world for what actually makes concepts what they are. That's quite a challenge, for there is a lot of confusion about the status of concepts. There are, in fact, many mistaken ideas about concepts. Here are some of them.

The most common, yet mistaken, idea is that concepts are thoughts. But that cannot be true. The reason is basically very simple: if concepts were just thoughts, then we would not be able to communicate with each other, because I cannot read your thoughts, and you can't read mine. Therefore, I wouldn't know what you are thinking when I use a certain concept to express what I am thinking. No one can tap into my brain and steal my thoughts, so I would never be able to know whether we are thinking the same thing when we speak about planets, protons, enzymes, genes, neurons, or whatever we wish to talk about.

Instead, when the two of us talk to each other, there must be something we have in common, something that transcends our private thoughts. That's where concepts come in. In other words, we may think about a concept, but the concept itself is not a thought—only the object of a thought. Concepts do not come from thoughts, but they make thoughts possible. According to St. Augustine, there must be something "that all reasoning beings, each one using his own reason or mind, see in common."[218]

Another mistaken idea is that concepts are merely linguistic entities. But that cannot be true either. It is true that words are linguistic items, but concepts are not—they are extra-linguistic entities. So we shouldn't confuse words with concepts. Words are, at best, labels for concepts. The word "protein" would be a label for the concept of "protein." A concept such as "protein" can be conveyed by different words in different languages—for example, the word "protein" in English and the word "*Eiweiß*" in German. However, the concept itself would exist even if those languages did not, or if they lacked the words to refer to the concept of "protein"—say, before the year 1800. We may talk about concepts with words, but the concept itself is not a word. Put differently, concepts do not come from language; instead, they make language possible.

Another mistaken idea is that concepts are mere labels for what words refer to—some extra-mental, material objects in our world. But that view ignores that the meaning and the reference of a word are two different things. The concept behind a certain word does not automatically or unambiguously refer to a particular physical object in the outside world. Without concepts, words don't refer to anything. Sometimes, the relation of reference is hoped for (e.g., the concept of "Higgs boson") or is imaginary (e.g., the concept of "centaur") or is stipulated (e.g., the concept of an irrational

[218] Augustine, *On the Free Choice of the Will* (Cambridge, UK: Cambridge University Press, 2010), 2.8.20.79.

number such as π). Concepts don't *create* reality—they try to *capture* it, and they may succeed or fail. As Stephen Barr put it, "One can have 4 cows, but one cannot have π cows; and one can have a 4-sided table, but not a π-sided table."[219] Yet we know the meaning of π, although it does not have any concrete reference in the physical world.

Another mistaken idea is that concepts come from definitions that we make up ourselves to explain what they stand for. But the problem of this idea is that definitions inevitably require other concepts. Take the following two definitions in a dictionary: "A gene can take alternative forms called alleles," and "An allele is one of the alternative forms of a gene." Obviously, these are two circular references; they explain a concept with the very same term it is supposed to explain. But even when the definition is not that trivial, but provides a more extensive description, the circularity remains, for every description of a concept requires the use of other concepts, which in turn require additional descriptions, and so on and so on.

Ultimately, dictionaries cannot avoid using circular references since all words in a dictionary are defined in terms of each other. Dictionaries can never step outside their own confines to refer to something beyond them. No matter how hard we try, we will never be able to get off the ground the concept we try to describe and define—every trial falls back on other concepts. Even if we decide to stop this endless regress by declaring a few concepts as the pillars that carry the rest of the conceptual framework, then we still need to explain where those presumably fundamental concepts come from.

Another mistaken idea is that concepts come directly from observation. That's hard to believe too. Concepts have no necessary

[219] Barr, *Modern Physics*, 194.

link to a particular situation. We can talk about a "mammoth" without ever having seen one. We can talk about "Atlantis" without ever having been able to locate it. We can talk about "π" without ever being able to find an example of it in the real world.

But what is even more crucial is this: observation is always about particular things, whereas concepts are universal. Concepts abstract from particular observations that which is universal. Concepts have the universality that observations miss. Material things are always particular, whereas concepts are always universal. We may have seen many circular objects in particular, but we have never seen the perfect, universal circle that the concept of "circle" describes.

Another mistaken idea is that concepts are established by pointing at something. First of all, for certain concepts, there may be nothing to point at. There is nothing to point at to explain the concept of "tomorrow," for instance (other than a calendar, but that requires the concept of calendar as well). Neither does the mathematical concept of π refer to any object in the world. It's not only universal, but also highly abstract. And the concept of "phlogiston" in early chemistry turned out to be pointing to nothing.

Second, as Wittgenstein explained, pointing at things is not sufficient to define words and their concepts.[220] Of course, I can explain what the word "red" means by pointing at a red tulip. But that gesture is very ambiguous. Perhaps someone else thinks "red" stands for a tulip, or for a flower, or whatever else might come to mind in connection with pointing at a red tulip. The word "red" is based on an abstraction—a concept, that is—of what we perceive. What pointing makes us see is the red object pointed at, but not the concept of redness.

Third, even pointing at several red objects cannot generate the concept of red. In order to decide which objects are to be included

[220] Ludwig Wittgenstein, *Philosophical Investigations*, §§258–277.

in the set of red objects and which are not, we need a criterion that says only "red" objects are to be included in the set. This account presupposes the very concept of red, the acquisition of which it is meant to explain. It's only when we know the concept of "red" that we can point at a red object and identify it as red. Many things can be pointed at once they have been identified, but not everything that has been pointed at can be identified with a concept. As St. Augustine once put it, "What is recognized is present in common to all who recognize it."[221] In other words, there must be some cognition before there is recognition.

Another mistaken idea is that concepts are material entities themselves, located *outside* the mind. But this idea cannot be true either. Even when a concept does refer to a material entity, the concept itself is not a material entity outside the mind in the world around us. Concepts don't have material properties—they have no weight, no shape, no color. Therefore, the concept of "circularity," for instance, is not circular itself; the concept of "electron" is not an electron itself; the concept of "gene" is not a gene itself. Particular entities in the world around us, outside our minds, are material, whereas concepts are immaterial and universal.

This has led some to another mistaken view of concepts: as material entities *inside* the brain. The famous example to defend this position is this: when a frog sees a fly zooming by, the frog's brain displays a certain pattern of neural firing.[222] In a similar way, when we see a tree, our thinking of a certain concept—say, a "fly" or a "tree"—must be only a certain pattern of neural firing.

However, the problem here is that seeing a fly or a tree is a matter of perception, whereas using a concept is a matter of thinking

[221] Augustine, *On the Free Choice*, 2.10.28.112.

[222] H. B. Barlow, "Summation and inhibition in the frog's retina," *Journal of Physiology* 119 (1953): 69–88.

and cognition. Using the philosopher Mortimer Adler's distinction again, conceptual content is not the same as perceptual content.[223] Conceptual content is about trees and flies in general, not about one particular tree or fly. It is universal, not particular; it is something mental, not neural. Thoughts and concepts have meaning and content, while perceptions do not.

Besides, to reduce concepts to a "product of neurons" or a "product of genes" obscures the fact that both "neuron" and "gene" are abstract concepts themselves. That would make for a pernicious vicious circle: the very idea that concepts are nothing but neuronal firing is itself nothing but neuronal firing. Those who claim that concepts are merely products of neurons or genes should realize that talking about neurons or genes requires the immaterial concept of neuron or gene to begin with. In other words, concepts do not come from neuronal activities or gene actions; instead, we can understand neuronal or gene activities only with the help of certain concepts such as "nerve," "neuron," "synapse," "neurotransmitter," and the like, or with the genetic concepts of "nucleotides," "transcription," "translation," "proteins," and the like.

It is hard *not* to conclude from all of this that concepts are very enigmatic entities. Thanks to concepts, we can see similarities that are not immediately visible and not directly tied to what we perceive. Everyone, even animals, can see things falling, but to perceive "gravity," one needs the concept of "gravity" in order to see what no one had been able to see before Isaac Newton. The concept of "gravity" allows us to "see," for example, the similarity between the motion of the moon and the fall of an apple. But that is conceptual content, not perceptual.

Put differently, a concept definitely goes beyond what the world shows us through our senses. We do not "see" gravity, because it

[223] Adler, *Intellect*, chap. 4.

is invisible. We do not "see" genes, but have come to hypothesize and conceptualize their existence and characteristics. We do not even see circles, for a "circle" is a highly abstract, idealized concept.

No wonder, then, that concepts play a central role in how we know the world. Concepts go far beyond what the senses provide—they transform "things" of the world into "objects" of knowledge, thus enabling us to see with our "mental eyes" what no physical eyes could ever see before. In short, concepts are cognitive tools. Take the mathematical concept of π again: it is not some private experience, such as a toothache; it is not a material object, such as a melon; it is more than a sensation, a neurological artifact, or a genetic product; it is certainly more than a certain pattern of neurons firing in the brain. But above all, it is not even a property of material things, because there are no entities in the physical world with the shape of π.

After all we have seen so far, we must come to the conclusion that concepts don't come from the animal world, nor from our thoughts, nor from our observations, nor from our languages, nor from our surroundings, nor from our brains with their neurons, nor from our genome with its genes. However, this conclusion immediately raises a crucial question: Where else, then, could they come from? If concepts are real, where do we find them? Where do they reside? They *must* exist somewhere for them to be available to each of us individually and all of us together. But where could that "somewhere" be?

Where do concepts come from?

Suddenly, we find ourselves in this strange, immaterial world of concepts, where things are not large or small, light or heavy, hard or soft. We can think about sizes and colors of things, but the thoughts themselves do not have sizes and colors—nor do the concepts they use. Amazingly, many scientists don't seem to be aware, at least not consciously, of the enigma that concepts present.

Fortunately, there are some welcome exceptions. The linguists Noam Chomsky and his co-author Robert Berwick are very honest about this issue when they admit, "In some completely unknown way, our ancestors developed human concepts."[224] Elsewhere, they say about concepts, "Their origin is entirely obscure, posing a very serious problem for the evolution of human cognitive capacities, language in particular."[225] Concepts turn out to be real enigmas.

If abstract objects such as concepts—and all the statements that contain them—do not depend for their existence on the material world or on the human mind, then there is only one rational option left: they must exist in a "third realm" that is neither material nor mental. This idea may sound outlandish, but it is certainly not new; it is usually associated with the Greek philosopher Plato, who mentioned this third realm long ago in his theory of forms.[226] However, Plato's position faces multiple, rather technical problems, which we will not discuss here.

But there is a much more acceptable version of this third realm—arguably the only valid one—which goes basically back to St. Augustine[227] and was later elaborated by the philosopher Gottfried Leibniz[228] and then strongly backed by the logician Gottlob Frege, who wrote in 1918 that "a third realm must be recognized"[229]—a realm of meaning or sense, independent of any-

[224] Berwick and Chomsky, *Why Only Us*, 87.

[225] Ibid., 90.

[226] Plato, *Timaeus*, 52.

[227] Especially so in book 2 of *On Free Choice of the Will*. See also *De Magistro* 11.36–37.

[228] Especially so in sections 43-46 of his *Monadology*.

[229] Gottlob Frege, "Thought," in *The Frege Reader*, ed. Michael Beaney (Oxford: Blackwell, 1997), 325–345.

one's particular ideas. In this version, abstract eternal entities or objects do indeed exist, but they can do so only in an *infinite, eternal Divine Intellect.*

This probably raises the question as to why abstract objects can't exist in *human* intellects, instead of one *Divine* Intellect. The reason is that human intellects are contingent—they do not have to exist, but come into being and pass away. If abstract objects existed only in human intellects, they would have to come into existence and could go out of existence again. Besides, we could not have them in common with other human intellects. And more importantly, before humanity emerged, there would and could not have been any abstract concepts—meaning snow wasn't white then. Of course, snow has always been white, but its whiteness became visible to us through the concept of "whiteness." Concepts are like light beams that hit a certain thing that is in darkness, such that this thing will now be in the light and become more visible. These concepts exist outside our minds in a world of their own.

Consequently, the only sort of intellect on which abstract, universal, and timeless concepts and statements could ultimately depend for their existence would be an intellect that could not possibly have *not* existed—a Divine Intellect, the Mind of God, the First Cause. It is this Divine Intellect that grasps and holds all of the logical relationships between all statements with all their universal concepts—an Intellect that eternally understands all actual truths and all possible truths, as well as all necessary truths. It is God who causes the world to be such that a statement is either true or false, either possible or impossible, either necessary or unnecessary. This is so because statements and their concepts exist as thoughts in the Divine Intellect.

To put it in a more charged way: without faith in God's Divine Intellect, we have nothing to claim as truth. We are entitled to say that the statement "snow is white" is true only if snow is indeed white in the Mind of God. One of the ways to find out whether something is indeed in the Intellect of God is by "reading God's Mind" in nature and through reason. We do so, for instance, when we "interrogate" the universe through investigation, exploration, and experiment, as well as when we use logic, reason, and philosophy.

How is it possible that reality, coming from the Divine Intellect, can ever be "grasped" by our human intellect? In some mysterious way, our human intellect is able to capture concepts and statements that reside in God's Divine Intellect. Only the Divine Intellect can make it possible that you and I share the same concept and statement when we say and think that snow is white or that mutations are random. Only the Divine Intellect can explain that the world is an objective and orderly entity knowable to the human intellect, which is also an orderly and objective product of the rational and consistent Divine Intellect. The book of Genesis could not have put it more simply: "God created man in his own image, in the image of God he created him."[230] God built us with the desire to know and learn and understand what exists in His Intellect.

Earlier, we mentioned the late astrophysicist Sir James Jeans, who once said, "The Universe begins to look more like a great thought than a great machine."[231] The "great thought" that Jeans speaks of here is not just a thought of the human intellect, but rather a "thought" of the Divine Intellect. The Divine Intellect is where concepts reside, but it's also where facts come from. Like concepts, facts, too, are peculiar nonmaterial entities.[232] A fact is not the same as an event: it is the *description* of an event, not the event itself. A

[230] Gen. 1:27.

[231] James Jeans, *The Mysterious Universe* (Cambridge, UK: Cambridge University Press, 1930), chap. 5.

[232] A. R. White, *Truth* (Garden City, NY: Doubleday, 1970), chap. 6.

fact is not the same as a statement; it is the *content* of a statement, not the statement itself. A fact is not the same as a thought; it is the *object* of a thought, not the thought itself. True, facts often need events so we can test them; they often need thoughts so we can understand them; and they often need statements so they can be discussed and shared by us. But if facts are neither events nor thoughts nor statements, then they can only exist in the Mind and Intellect of God. That's where their truth must originate.

Remarkably, one of the twentieth century's greatest philosophers of science, Karl Popper, once wrote, "Knowledge in the objective sense is *knowledge without a knower*; it is *knowledge without a knowing subject*."[233] Popper was right when it comes to the knowledge of individual people, whether they are scientists or not. But he was wrong in another sense: even without a human knower or a knowing subject, objective knowledge remains objective and thus must exist somewhere outside the human mind and prior to it—that is, in the Divine Intellect.

In other words, God is the "knower," the "knowing subject," behind all our human knowledge. There cannot be any knowledge without the knowing subject of God. Only this can explain why scientists can do their work of exploring and investigating nature. What they are actually doing through their research is reading the Mind of God—often without their knowing it, let alone acknowledging it. Stephen Hawking once put it this way: finding a unified theory in physics would be "the ultimate triumph of human reason—for then we would know the mind of God."[234] Indeed, God has given us the tools to read His Mind and to begin to think a bit more like Him.

[233] Karl Popper, *Objective Knowledge: An Evolutionary Approach* (London: Oxford University Press, 1972), 109. Emphasis original.

[234] Hawking, *Brief History of Time*, 175.

What may we conclude from this? Once we see a strong connection between God and the universe He created, between God's Intellect and what our human intellect knows about the world through concepts and statements, the universe becomes more intelligible and sensible. Semantics cannot dodge the question of where the meaning of words ultimately comes from. Apparently, even the use of concepts points to God.

11

How Logic and Math Point to God

Logic is the study of the methods and principles used in distinguishing correct from incorrect reasoning. Of course, one can reason correctly without having studied logic, just as one can walk well without having studied physics or physiology. Yet people who have studied logic are more likely to reason correctly and are better able to apply methods for testing the correctness of their reasoning. In reasoning, we want to make sure that asserting certain premises to be true warrants asserting the conclusion to be true too. We may call something "true" after observation or experimentation, but there is more to it: logic, that is, which makes us *conclude* it's true.

Mathematics is the study of topics such as quantity (number theory), structure (algebra), and space (geometry). It allows us to establish truth by rigorous deduction from appropriately chosen axioms and definitions—a so-called axiomatic system. Proofs in mathematics may seem to be safe and foolproof, but that's only true in a rather "artificial" way. Mathematical knowledge may be considered the most secure form of knowledge, but it is not about anything material. Albert Einstein said it well: "As far as the laws of mathematics refer to reality, they are not certain; and as far as they are certain, they do not refer to reality."[235]

[235] Address to Prussian Academy of Sciences, 1921.

Besides, even in mathematics, we cannot prove anything from nothing, for at least something is needed—axioms—to start the process. Based on a set of axioms in the premises, we can derive a conclusion with final certainty. Based on such axioms, we can conclude, for example, that the sum of the three angles of a triangle is 180 degrees. So the certainty of the conclusion is based on the truth of the axioms. That's quite a trivial way of proving things. Yet it still offers more certainty than we can ever reach in science.

Before we get "too certain" about the geometry we learned in school, Euclid's geometry, it needs to be said that there are other geometries with a different set of axioms. What is true in one of these geometries may be false in the others. In Euclidean geometry, one axiom says that one line, and only one, can be drawn parallel to a given line through a point not on the line. These two lines, according to the so-called parallel axiom, remain at a constant distance from each other.

Euclid had postulated only one parallel line through a point not on the line. But that is not the case for geometry à la Lobachevsky, where these lines "curve away" from each other, so there can be *many* lines parallel to a given line through a point not on it. As a consequence, the sum of the angles of a triangle projected in a hyperbolic space is *less* than 180 degrees. And then there is another geometry à la Riemann, where the lines "curve toward" each other and intersect, so there is *no* line parallel to a given line through a point not on it. As a consequence, the sum of the angles of a triangle projected in an elliptic space is *more* than 180 degrees.

Is one of these geometries better than the others? Yes and no. For daily life, Euclid's geometry works fine. But during the nineteenth century, several mathematicians realized that a geometry based on Euclid's axioms was not the only possibility. A non-Euclidean universe was too counterintuitive for many to accept at first. Yet it was non-Euclidean geometry that paved the way for Albert Einstein's theory of general relativity in the early 1900s and the modern understanding of space-time.

Geometric curvature is hard to imagine for Euclidean minds. But think of a sheet of rubber with grid lines on it, suspended horizontally as a flat surface. With no weight on it, the grid has straight lines and right angles, corresponding to the "flat space" of Euclidean geometry. But when you place a ball on the surface, the rubber sheet stretches around it. The curvature of the grid increases as it gets closer to the ball. This corresponds to the curvature of space-time near a massive object.

What logic and math have in common

What logic and mathematics have in common is that, unlike science, they are not about the "real" world—they can be done from an armchair. We can use them when dealing with the real world, but they don't directly refer to it. This can be shown with a very simple example. When we add one to one in math, the result is necessarily two (so $1 + 1 = 2$); but when we add one drop of water to another, the result is not two drops but one drop (so $1 + 1 = 1$); and when we add one organism to another, we may end up with three or more of them (so $1 + 1 \geq 3$).

This explains why logic and mathematics are in essence about nothing—or perhaps, in a different sense, about everything. The best place to show this is probably mathematics. All the axioms and conclusions mathematicians use about a circle, for instance, are not about real circles, for there are no real circles in this world. They are about something very ethereal: a perfect circle. But perfect circles don't exist. Some circles we know may come close to a perfect circle, but that's where the similarity ends.

In logic, we find something similar. The logical rule that nothing can contradict itself is not about any particular contradiction we hear or read about, but rather about contradictions in general. Or take the rule of inference called *modus ponens*, which takes two premises, one in the form of "if *p*, then *q*" and another in the form of "*p*," and then returns the conclusion "*q*." This rule is true for any *p* and any *q*. The same can be said about other rules of inference such as the *modus tollens*, which takes two premises, one in the form of "if *p*, then *q*" and another in the form of "not *q*," and then returns the conclusion "not *p*." This is basically what happens when a hypothesis *p* is being falsified.

What logic and math have in common, too, is that they can lead us on a safe path. Thanks to the laws of logic and math, we are able to argue, debate, reason, prove, and disprove. For instance, logic tells us that objects that came into being cannot explain their own existence, for they would have to exist before they came into existence, which is logically impossible. Or one could argue, with the biologist J. B. S. Haldane, that "if materialism is true, it seems to me that we cannot know that it is true. If my opinions are the result of chemical processes going on in my brain, they are determined by chemistry, not the laws of logic."[236] That's something logic can demonstrate for us with certainty.

Something similar holds for math. Using mathematical rules, we can derive safe conclusions. A simple example would be this: $(-3)^2$ equals 9 [because $(-3)(-3) = 9$]. In other words, $(-3)^2$ does not equal -9 [based on $-(3)(3) = -9$], for that would be a wrong conclusion in mathematics. Or take a triangle in geometry. Deriving from the properties of a triangle based on Euclid's axioms, we conclude that the sum of its three angles must be 180 degrees—any other conclusion would be false, unless we use different axioms, as is the case, for instance, in Riemann's geometry.

236 J. B. S. Haldane, *The Inequality of Man* (New York: Penguin Books, 1932), 162.

However, logic and math can also mislead us. This is not to say that logic or math is at fault—we are. If a deductive argument leads us to faulty conclusions, then we cannot blame logic. We may have reason not to accept the premises, so the conclusion does not follow. Or there may have been some *hidden* premises or assumptions involved that we were not aware of or had forgotten to take into consideration. They can put the conclusion of a seemingly safe deductive argument in jeopardy.

As an example of this latter possibility, we could mention the explorer Christopher Columbus. He reasoned in a deductive way that the earth must be round as follows: as a ship sails away from shore, its upper portions remain visible to an observer on land long after its lower parts have disappeared from view—so the earth must be round. But his logic was not valid because of a hidden premise, which states that light rays follow a rectilinear path. If they followed a curved path, concave upwards, we would still see the same happening to the ship, even on a flat earth! Perhaps some of Columbus's sailors were still afraid to fall off the edge of the flat world!

Mathematics is not the final truth either. Mathematical structure by itself is a mere abstraction; it cannot be all there is to math, because structure presupposes something concrete that *has* the structure. Albert Einstein, for one, always resented a growing trend of "mathematicalization" in physics. He did not want to be a slave to mathematics. For instance, after he had studied Fr. Lemaître's revolutionary 1927 paper intensely, he told the priest in person that his mathematics was perfect, but his physics abominable. Einstein would one day take those words back, but his point was that mathematics does not and should not have the last word in science.

To make a long story short, there are many similarities between logic and math. They are actually so close that scholars have tried to "translate" mathematical issues into logical issues. In the nineteenth century, mathematicians became aware of some serious logical gaps and inconsistencies in their field. It was shown that Euclid's axioms for geometry, which had been taught for centuries as the ideal example of the axiomatic method, were at least incomplete.

As a result, mathematicians began to search for axiomatic systems that could be used to formalize large parts of mathematics. One of the earliest examples was the trendsetting book *Principia Mathematica*, authored by Alfred North Whitehead and Bertrand Russell.[237] They tried in this book to express mathematical propositions precisely in terms of symbolic logic. They defended the view that all mathematical truths are logical truths, and that mathematics is only a chapter of logic. Believe it or not, the authors had to go a long, long way to prove the validity of the proposition 1 + 1 = 2. No wonder the physicist Eugene Wigner says it's a "miracle that the human mind can string a thousand arguments together without getting itself into contradictions."[238]

The *Principia* covered only set theory, cardinal numbers, ordinal numbers, and real numbers, but by the end of the third volume, it was clear to experts that a large amount of known mathematics could in principle be developed in the formalism of logic. It was also clear how lengthy such a development would be. Although a fourth volume on the foundations of geometry had been planned, the authors admitted to intellectual exhaustion upon completion of the third.

[237] Alfred North Whitehead and Bertrand Russell, *Principia Mathematica* (Cambridge, UK: Cambridge University Press, 1920).

[238] Eugene Wigner, "The Unreasonable Effectiveness of Mathematics in the Natural Sciences," *Communications in Pure and Applied Mathematics* 13, no. I (February 1960).

The problem of contradictions

Both logic and math reject contradictions. To find out why, we need to distinguish at least three different kinds of contradictions.[239]

First, there are contradictions about the real world. Bertrand Russell gave us a simple example.[240] Suppose that in some isolated village, there is only one barber. Some people go to that barber, while some cut their own hair. The village has the following rule that is strictly enforced: everyone must get his hair cut; furthermore, if you cut your own hair, then you do not go to the barber; and if you do not cut your own hair, you go to the barber—it's one or the other.

But now ask yourself the question: Who cuts the barber's hair? If the barber cuts his own hair, then, because of the rule, he cannot go to the barber—but he is the barber! If, on the other hand, the barber goes to the barber, then he is cutting his own hair, in violation of the rule. These two results present a contradiction. The contradiction is that the barber must cut his own hair and at the same time must not cut his own hair. It's an example of a paradox.

What's wrong with contradictions? A contradiction is a statement or proposition that is both true and false. The fact is, though, that there can be no contradictions in the physical world—a thing cannot both be and not be, and a proposition cannot be both true and untrue. A paradox shows us that a given assumption cannot be true if a contradiction or falsehood can be derived from it.

Of course, there are simple solutions to the barber's paradox: the barber is bald, or the barber's wife cuts his hair, or the barber

[239] Noson Yanofsky, *The Outer Limits of Reason: What Science, Mathematics, and Logic Cannot Tell Us* (Cambridge, MA: MIT Press, 2016).

[240] Bertrand Russell, "The Philosophy of Logical Atomism," in *The Collected Papers of Bertrand Russell, 1914–19*, vol. 8., 228.

quits his job as a barber, and so on. In fact, these ideas are shadows of the true resolution to the barber's paradox, which is to realize that the village with this strict rule just cannot exist. In short, the assumption of the barber's paradox was that the village with this rule does exist—which is a false assumption. Once we abandon this assumption, we are free of the contradiction. So such a paradox shows only that the physical world has limitations.

Second, there are contradictions that arise from language. The most famous paradox in language is the so-called liar's paradox. The simple statement "This statement is false" is apparently both true and false. If it is true, then it asserts that it is false; on the other hand, if it is false, then it must be true since it claims to be false. Because the statement is both true and false, it's a contradiction.

What can be done about the liar's paradox? Some philosophers say that the statement "This statement is false" is actually not declaring anything about anything. Such statements are literally nonsensical. Rather than trying to resolve the paradox by avoiding the contradiction, they ignore the paradox as nonsensical language, simply saying that the statement is neither true nor false.

Third, there are contradictions that arise in logic and math. Mathematics is a language that human beings use to make sense of the world. Since math is a language, it can potentially have contradictions. However, because we want to use mathematics in science to talk about the physical universe, which cannot have contradictions, we must make sure that the language of mathematics does not have contradictions either, as we must ensure that our scientific predictions about the physical world don't lead to contradictory conclusions.

The first time most of us encountered a contradiction in mathematics was probably when we were informed that we can divide any number by any other number . . . except zero. Why that exception? If we divide by zero, we can derive a contradiction. Consider the true statement $0 \times 2 = 0 \times 3$ (a lot of nothing is still nothing).

If we divide both sides by zero to cancel out the zeros, we are left with $2 = 3$. So by dividing by zero we got a contradiction: $2 \neq 2$ ($\neq$ means "is not"). In short, paradoxes in mathematics show us that although we can perform certain operations, we should not, so we don't create a contradiction.

Bertrand Russell developed another paradox, this time concerning so-called classes or sets.[241] Sets may be divided into two different groups: normal sets, which do not contain themselves as members, and non-normal sets. which do. Now let *N* stand for the set of all normal sets. Then, we ask ourselves whether *N* itself is a normal set. If so, it is a member of itself. But in that case *N* is non-normal, because by definition a set that contains itself is non-normal. Yet if *N* is non-normal and thus a member of itself, it is must be normal, because by definition all the members of *N* are normal. In short, *N* is normal if and only if *N* is non-normal—a great contradiction! From this, Russell drew the conclusion that there *is* no such thing as a set of *all* sets, for the notion of a set of all sets that includes itself as a member generates contradiction (just as there is no town like the one depicted in the barber's paradox).

So where did we end up? All paradoxes tell us there is something wrong with what they reveal. The paradoxes about reality show us that the physical world has its own limitations. The paradoxes arising from logic and math show the limitations of reason. The paradoxes of language are merely a result of sloppy language use.

What do logic and math have to do with God?

Let's start with the paradoxes we just discussed. What is wrong with contradictions? Why should we not accept them? Many answers

[241] Bertrand Russell, "Logic and Knowledge," *American Journal of Mathematics* 30 (1908): 222–262.

can be given: it's not fair play, it undermines everything we think we know, it would be irrational, we could conclude anything we want, and so on. But those cannot be final answers, for we could just decide to accept contradictions as part of normal life.

A better answer would be that the world was created by God in such a way that contradictions should not be found anywhere. Nothing in this world can both be and not be. Could God have created a world that harbors contradictions? No, that would be impossible because that's a contradiction in itself. God is a God of Reason, so He is not able to do what is logically contradictory, according to St. Thomas Aquinas. Aquinas gives many examples of what God cannot do: God cannot create square circles; God cannot create triangles with four sides; God cannot declare true what is false; God cannot declare right what is wrong; God cannot undo something that happened in the past; and the list goes on and on.[242] To do such things would be contradictory, and therefore against reason, and therefore impossible in God's creation. Whatever is contradictory in logic and math is also contradictory in God's Mind. Aquinas concludes from this, "Hence it is better to say that such things cannot be done, than that God cannot do them."[243]

Is God really needed here to explain why contradictions must be avoided? Isn't there another explanation for the fact that contradictions are impossible in logic and math? Some think this explanation can be found in biology. They call upon Darwin's process of natural selection to explain how our mind acquired logic and math. However, it's hard to defend that natural selection led our reasoning power in logic and math to the perfection that it definitely seems to possess. Natural selection does not favor belief

[242] Aquinas, *Summa Contra Gentiles* II, 25.

[243] Aquinas, *Summa Theologica* I, q. 25, art. 3.

in prime numbers and perfect triangles, for instance. There is no survival-value in such beliefs.

Nevertheless, good and bad logic and good and bad math definitely exist. As G.K. Chesterton liked to ask his readers, "Why should not good logic be as misleading as bad logic? They are both movements in the brain of a bewildered ape?"[244] Yet we know good logic or math does not mislead us the way bad logic or math does. In fact, science heavily relies on good logic and mathematics to interpret consistently the data that scientific observation and experimentation provide. And the history of science shows its success and what we have gained by using good math and good logic.

Apparently, logic and math are not so formal that they are completely irrelevant for the real world. The famous logician Gottlob Frege, for instance, always strongly defended the scientific status of logic and mathematics against the claim that their laws are mere psychological principles governing the operation of the human mind.[245] According to the latter claim, they don't describe objective reality, but merely the structure of our minds in thinking about reality. However, this one-sided view raises the problem of how we *know* this view is true. As Edward Feser remarks, in order for us to defend this position, it must have some foundation in objective reality—which is ruled out by this very position ahead of time.[246] That makes for another case of contradiction and self-destruction.

Physics may have become heavily "mathematicalized," as we mentioned earlier, but its equations do retain a strong link to the physical world; otherwise, they would have to be corrected or abandoned. Take the simple mathematical expression of the law of

[244] G.K. Chesterton, *Orthodoxy* (Chicago: Moody Publishers, 2009), 33.

[245] Frege, "Thought."

[246] Edward Feser, *Five Proofs of the Existence of God* (San Francisco: Ignatius Press, 2017), 96.

gravity—$(m_1 \times m_2) / r^2$. This formal rendition is definitely about reality. What can be more "real" than the mass of two objects [m_1 and m_2] and the distance between the centers of the two masses [r^2]? There is no way this could be just a construction of the human mind. Something similar could be said about geometry. Euclid's system works pretty well on planet earth, but Riemann's system works better on a much larger scale. However, that doesn't make Euclidean geometry useless.

So there must be some explanation for the fact that the human intellect, with its logic and math, seems to be able to capture reality the way it *is*. Not only does the physical order we observe in this world appear to be amazingly consistent, but so does the order of logic and math. Somehow, the consistency of logic and math reflects the consistency of the world. It is a consistency that must perplex us. It calls for an explanation of what the physicist Eugene Wigner calls the "unreasonable effectiveness of mathematics."[247] This explanation must be found in the rationality of the human mind, with its logic and math, as being a reflection of the rationality that was implemented by God, the Creator, in His creation. That's definitely another pointer to God.

Let's address one more question: What makes logical and mathematical truths *necessarily* true? If we answer with logic or mathematics, then we end up in a form of circular reasoning. How can we step out of this circle? The only possible answer is that their truth must reside in the eternal world of God's Mind and Intellect. It is thanks to God that thirteen is a prime number in math and that nothing can contradict itself in logic. Even the geometries of Euclid, Riemann, and Lobachevsky are in God's Mind and Intellect, and that's how they can be in our minds as well.

[247] Wigner, "Unreasonable Effectiveness."

The eminent logician Gottlob Frege strongly agrees with this thought: "Thus for example the thought we have expressed in the Pythagorean Theorem is timelessly true, true independently of whether anyone takes it to be true. It needs no owner. It is not true only from the time when it is discovered; just as a planet, even before anyone saw it, was in interaction with other planets."[248] The theorem is eternally true in the Mind of God—and that's what makes this theorem and others possible, and perhaps factual.

After all we have seen in this chapter, it is hard to avoid the conclusion that the rationality and consistency found in logic and math reflect the rationality and consistency found in the Divine Intellect. That's how logic and math carry many pointers to God.

[248] Frege, "Thought," 337.

12

How Gödel's Theorem Points to God

In the previous chapter, we defended the view that the whole of mathematics and logic should be proved to be consistent and free from any contradictions. A system is called consistent if no contradictions can arise in it, so that it is impossible to derive two contradictory statements from it, such as both "*q*" and "*not q*." But this proof turned out to be difficult for several theorems, and it may even turn out to be an impossible dream anyway.

Something is missing

A theorem is a statement that has been proven on the basis of previously established statements, such as other theorems, or generally accepted statements, such as axioms. A theorem is supposed to be a logical consequence of axioms.

Pythagoras's theorem is probably the best known example of a theorem. It is about a fundamental relation in Euclidean geometry among the three sides of a right triangle. It relates the length of a triangle's hypotenuse (c) to the lengths of the triangle's other two sides (a and b). It states that the area of the square whose longest side is the hypotenuse (the side opposite the right angle) is equal to the sum of the areas of the squares on the other two sides. This theorem can be written with the following equation: $a^2 + b^2 = c^2$. It has at least 370 known proofs, either algebraic or geometric.

However, not all theorems have been backed by proofs. A classic illustration of a mathematical theorem that defies all attempts at proof is the statement that every even number is the sum of two prime numbers. Although no one has found an even number that is not the sum of two primes, no one has ever succeeded in finding a proof that this applies to *all* even numbers. Searching by trial and error does not work here, as we need a proof for *all* even numbers, but that proof has never been given.

In the previous chapter, we discussed how the authors of *Principia Mathematica* thought they had developed a large amount of known mathematics with the formalism of logic—enough to avoid contradictions and inconsistencies. However, it came quite as a blow when Bertrand Russell showed with his famous *set paradox* that contradictions were not precluded by such formal approaches. From this, Russell drew the conclusion that there *is* no such thing as a set of *all* sets, for the notion of a set of all sets that includes itself as a member generates contradiction. The concept of "set" may seem crystal clear, but it isn't on further inspection.

Deriving contradictions from a formal system is supposed to be lethal. So this possibility has to be excluded in a consistent formalized axiomatic system. How can this be done? One could check first that the rules of the system never allow anyone to derive a contradiction from them. But then one needs to check whether the methods used to prove that the system is consistent are consistent themselves. And one could go on and on, which amounts to endless regress.

But there is, in the words of Stephen Barr, a possible solution: "using only methods of reasoning that are available within that formal system. Then one will have both proven the system's consistency and at the same time shown that the steps of reasoning used to do this were themselves consistent. All doubts about consistency would then be laid to rest forever."[249]

This became known as Hilbert's program of providing a consistency proof of arithmetic within any formal theory of arithmetic. The main goal of Hilbert's program was to provide secure foundations for all mathematics (and logic). Let's leave it at that.

There is always something missing

It has been known for a while that there is something missing in several widely accepted theorems. But how can we know that there is *always* something missing in whatever theorem we want to prove? It was Kurt Gödel—born in Brünn, Austria-Hungary in 1906 (now Brno in the Czech Republic)—who rigorously showed that all efforts to prove strictly that the whole of mathematics and logic is consistent and free from any contradictions are doomed to failure. He did so in what became known as the *incompleteness theorem*. In fact, Gödel published two incompleteness theorems in 1931 when he was twenty-five years old, one year after finishing his doctorate at the University of Vienna.

Gödel's proof of the incompleteness theorem is extremely difficult to read. A reader must master forty-six preliminary definitions, together with several important preliminary theorems, before getting the main results. I will try to keep my summary as brief and simple as possible, for it's a monumental proof.[250]

In mathematics, any provable statement must by definition be true, because "proof" means "mathematically demonstrated to be true." Before Kurt Gödel came along, mathematicians assumed

[249] Barr, *Modern Physics*, 281.

[250] Ernest Nagel and James R. Newman, *Gödel's Proof* (New York: New York University Press, 2008), 68–108.

that any mathematical statement that is true is also provable. In his first incompleteness theorem, Gödel showed this assumption is false by formulating a mathematical statement similar to the liar's paradox, which essentially says of itself, "This mathematical statement is unprovable."

Let's think about that for a moment. If the mathematical statement "This mathematical statement is unprovable" is indeed unprovable, then it is true. Alternatively, if "This mathematical statement is unprovable" is provable, then it is false. But how can proving it lead us to a false statement? That amounts to a contradiction, since all mathematical proofs are by definition proofs of true statements. So we must conclude that this statement is both true and unprovable. Such a statement creates a sort of no-win situation: if it is provable, then it is false, and the system is therefore *inconsistent*; and if it is not provable, then it is true, and the system is therefore *incomplete*. A paradox has led us to a restriction of the power of mathematics: there are statements in mathematics that are true, but can never be proven.

Gödel's second incompleteness theorem shows that no formal system of arithmetic can be used to prove its own consistency. Thus, the statement "There are no contradictions in the *Principia* system," for instance, cannot be proven in the *Principia* system unless there are contradictions in the system—in which case it can be proven to be both true and false.

In short, Gödel proved two things: if a logical system is consistent, it cannot be complete; and the consistency of axioms cannot be proved within their own system. Thus there will always be at least one true but unprovable statement. That means, for any set of arithmetic axioms, there is a formula that is true of any axiomatic system, but that is not provable in that system. In other words, any complete formal system is inconsistent, while any consistent formal system is incomplete—in short, there is always something that is missing.

What is missing?

If no system, no matter how formalized and comprehensive, is complete, then the question arises as to what is missing. As Gödel showed us, no coherent system—not even the system of science—can be completely closed; any coherent system is essentially incomplete and needs additional "help" from outside the system, according to Gödel. His theorem proves that no system can ever fill its own gaps. To accept the consistency of the system requires an "act of faith."

Gödel even went as far as believing that a credible account of reality itself cannot be completely closed and would therefore be incomplete if we do not invoke help from above, from God.[251] Gödel was said to be very cautious in mentioning this belief in scientific circles, because he considered it potential dynamite. But what he did say is that our capacity to know truth transcends mere formal logic; in other words, there are truths that we cannot "prove" in a logical way. The truth of God's existence may be one of them, although philosophers such as St. Thomas Aquinas claim there are, in fact, proofs of God's existence.[252] Interestingly, in the early 1970s, Gödel circulated among his friends at Princeton an elaboration of Leibniz's version of Anselm's ontological proof of God's existence. This is now known as Gödel's ontological proof.[253]

Some people think that what we've said here is reading too much into Gödel's proof and his religious convictions. I don't believe that's true. Fear of ridicule and professional isolation made him reluctant to talk about his faith. "Ninety percent of

[251] David Goldman, "The God of the Mathematicians," *First Things* 205 (April 2010): 45–50.

[252] Verschuuren, *Catholic Scientist Proves God Exists*.

[253] Hao Wang, *A Logical Journey: From Gödel to Philosophy* (Cambridge, MA: MIT Press, 1996), 316.

contemporary philosophers see their principal task to be that of beating religion out of men's heads," he wrote to his mother in 1961.[254] His biographer Rebecca Goldstein, who was a graduate student at Princeton during Gödel's last years there, testifies to this, snickering that she and her peers "found it hilarious" that Gödel "deluded himself into believing that God's existence could be proved a priori."[255] That's how intolerant the scientific community has become. Ironically, Gödel's religious insights may have been challenged by some, but his incompleteness theorem has never been called into question.

It was at Princeton, too, that Kurt Gödel and Albert Einstein developed a strong friendship. They were known to take long walks together to and from the Institute for Advanced Study. These two giants, one of mathematics and the other of physics, had a lot in common to talk about. According to the physicist Freeman Dyson, who worked at the same institute, Gödel was "the only one who walked and talked on equal terms with Einstein."[256]

They could certainly talk extensively together during their walks to and from the institute, but their religious beliefs were different. While the God of Einstein was Spinoza's God, the God of Gödel was more in line with Leibniz's. As Gödel himself wrote, "My belief is theistic, not pantheistic, following Leibniz rather than Spinoza."[257] He also told the late logician and mathematician Hao Wang of Rockefeller University, who chronicled Gödel's philosophical ideas, that "Spinoza's god is less than a person;

[254] Kurt Gödel, *Collected Works*, vol. 4 (Oxford: Clarendon Press, 2003), 436–437.

[255] Rebecca Goldstein, *Incompleteness: The Proof and Paradox of Kurt Gödel* (New York: W. W. Norton, 2006), 210.

[256] Freeman Dyson, *From Eros to Gaia* (New York: Penguin Books, 1993), 16.

[257] Wang, *Logical Journey*, 112.

mine is more than a person; because God can play the role of a person."[258]

It is hard to deny that Gödel, who was brought up in the Lutheran faith, was indeed a convinced theist. According to his wife, Adele, "Gödel, although he did not go to church, was religious and read the Bible in bed every Sunday morning."[259] This most likely explains how Gödel could see his incompleteness theorem as pointing to God. Only God can make everything complete, so there is no longer anything missing. Seeing the world in the light of God as well as from the perspective of science is something the English theoretical physicist and Anglican priest John Polkinghorne describes as seeing the world with "two eyes instead of one." As he explains, "Seeing the world with two eyes—having binocular vision—enables me to understand more than I could with either eye on its own."[260]

[258] Ibid., 152.

[259] Ibid., 51.

[260] John Polkinghorne, "God vs. Science," interview by Dr. Dean Nelson, *Saturday Evening Post*, August 20, 2011.

Conclusion

It is not uncommon for people to close their eyes and ears to evidence, even when it's overwhelming evidence—this happens all the time in court, for instance. But it is hard to keep ignoring or dismissing evidence, for it keeps asking for attention. St. Paul clearly noticed all that pointed to God at a time when there was hardly any science.[261] But now that science has taught us so much more, we have discovered even more evidence that keeps stubbornly pointing to God.

In this book, we found out how much in various fields of science points to God. They are pointers, though, not proofs. But all of them combined make it hard to deny that there is a God who made all of this possible, including science. Denying God's existence would actually be detrimental to everything science has done and will do in the future. It would, in fact, undermine all scientific activities and claims.

From this it follows that putting science in conflict with religion is a no-no. They are not exclusive alternatives. To use a silly example, we don't have to decide whether the Grand Canyon was carved by God or by the Colorado River. That's an absurd contrast.

[261] Rom. 1:20.

The Grand Canyon was made by rivers and streams—secondary causes—yet it was created by God—the First Cause. Without God, there could be no rivers or streams, actually no Grand Canyon at all. In a similar way, it is absurd to say that children cannot come from God because they come from their parents, or that children cannot come from their parents because they come from God. To put science and religion in contrast to each other would be equally absurd.

Although science can certainly do a lot for us, it has at least two caveats. On the one hand, what science can do for us is of a limited scope—restricted to what can be dissected, measured, and counted. But science is not the only way of knowing. On the other hand, as we found out, science could never live without religion, because science works with assumptions that it could never validate on its own, without the help of religion—assumptions about reality, causality, intelligibility, order, and experimentation. In other words, science can do a lot for us only if it acknowledges that what it can do is possible only because of what it cannot do—namely, testing and proving its own assumptions, which unmistakably point to God.

The truth of the matter is that science and religion can coexist: they have coexisted, they coexist right now, and they will continue to coexist in the future. They can certainly live together in the same "town." They have the same goal of explaining reality—whether its material or its immaterial aspects. Science and religion are the two "windows" that let humans look at the world they live in. Yet there is only one world, and it is the same world these two windows try to reveal to us. Closing off either window would make us partially blind.

Therefore, we must conclude that science and religion not only are able to live together, but also *must* live together.

Index

About the Author

Gerard M. Verschuuren is a human biologist, specializing in human genetics. He also holds a doctorate in the philosophy of science. He studied and worked at universities in Europe and in the United States. Currently semiretired, he spends most of his time as a writer, speaker, and consultant on the interface of science and religion, faith and reason.

Some of his most recent books are: *In the Beginning: A Catholic Scientist Explains How God Made Earth Our Home*; *Forty Anti-Catholic Lies: A Mythbusting Apologist Sets the Record Straight*; *Darwin's Philosophical Legacy: The Good and the Not-So-Good*; *God and Evolution?: Science Meets Faith*; *The Destiny of the Universe: In Pursuit of the Great Unknown*; *It's All in the Genes!: Really?*; *Life's Journey: A Guide from Conception to Growing Up, Growing Old, and Natural Death*; *Aquinas and Modern Science: A New Synthesis of Faith and Reason*; *Faith and Reason: The Cradle of Truth*; *The Myth of an Anti-Science Church: Galileo, Darwin, Teilhard, Hawking, Dawkins*; and *At the Dawn of Humanity: The First Humans*.

For more information, visit https://en.wikipedia.org/wiki/Gerard_Verschuuren.

Verschuuren can be contacted at www.where-do-we-come-from.com.

Sophia Institute

Sophia Institute is a nonprofit institution that seeks to nurture the spiritual, moral, and cultural life of souls and to spread the Gospel of Christ in conformity with the authentic teachings of the Roman Catholic Church.

Sophia Institute Press fulfills this mission by offering translations, reprints, and new publications that afford readers a rich source of the enduring wisdom of mankind.

Sophia Institute also operates the popular online resource CatholicExchange.com. *Catholic Exchange* provides world news from a Catholic perspective as well as daily devotionals and articles that will help readers to grow in holiness and live a life consistent with the teachings of the Church.

In 2013, Sophia Institute launched Sophia Institute for Teachers to renew and rebuild Catholic culture through service to Catholic education. With the goal of nurturing the spiritual, moral, and cultural life of souls, and an abiding respect for the role and work of teachers, we strive to provide materials and programs that are at once enlightening to the mind and ennobling to the heart; faithful and complete, as well as useful and practical.

Sophia Institute gratefully recognizes the Solidarity Association for preserving and encouraging the growth of our apostolate over the course of many years. Without their generous and timely support, this book would not be in your hands.

www.SophiaInstitute.com
www.CatholicExchange.com
www.SophiaInstituteforTeachers.org

Sophia Institute Press® is a registered trademark of Sophia Institute.
Sophia Institute is a tax-exempt institution as defined by the Internal Revenue Code, Section 501(c)(3). Tax ID 22-2548708.